ICE TIME

ICE TIME

CLIMATE, SCIENCE, AND LIFE ON EARTH

THOMAS LEVENSON

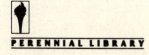

HARPER & ROW, PUBLISHERS, New York

Grand Rapids, Philadelphia, St. Louis, San Francisco
London, Singapore, Sydney, Tokyo, Toronto

First PERENNIAL LIBRARY edition published 1990.

Designed by: Sidney Feinberg

The Library of Congress has catalogued the hardcover edition as follows:

Levenson, Thomas.
  Ice time.

  Includes index.
  1. Climatology.   I. Title.
QC981.L48   1989   551.6   88-43002
ISBN 0-06-016063-2

ISBN 0-06-091703-2(pbk.)
90 91 92 93 94   **FG**   10 9 8 7 6 5 4 3 2 1

For my mother; with the memory of my father

Contents

Acknowledgments

A great number of people helped me throughout the writing of this book, and to all of them, my thanks. Some performed so far above and beyond the call of duty that I must mention them by name. First among them is my agent, Sallie Gouverneur, who shepherded me and my work; without her the book would never have been written. Richard Kot, my editor at Harper & Row, has enabled me to express my meaning far more clearly than I could ever have done unassisted.

Many people within the world of climate science gave extremely generously of their time; many more contributed indirectly, through their writing and their ideas—each mention of them in the text that follows is an expression of my thanks. Extra acknowledgment, however, must go to Stephen Schneider, who read the manuscript at several stages of preparation, served as my host at the National Center for Atmospheric Research (NCAR), and has over the years deepened my understanding of his subject. Cheryl Schneider read the manuscript and made valuable suggestions. Mary Rickel made my stay at NCAR a pleasure. Every member of the NCAR staff with whom I had any contact—too numerous to name here—made me feel welcome, and allowed me to depart well informed.

Darin Toohey read the entire book twice and worked very hard to keep me on the straight and narrow. Robert Boyle, Michael Oppenheimer, and George Woodwell provided me with information and encouragement.

My colleagues at NOVA have given me feedback and material support for the completion of the book. My thanks to Paula Apsell, William Grant, and the entire NOVA staff. David Kuhn of WGBH helped launch the project, and Gil Rogin at *Discover* magazine commissioned the articles that first placed me on the path that lead to this book.

In the course of writing, several friends and family members provided amazing amounts of support and interest sustained in the face of monomania; my deepest thanks to Merry White, Stephen Latham, David Ossman, Judith Walcott, Kelly Roney, Jon Eckstein, Michael Kosowsky, Jennifer Drawbridge, Diane Schacter, Edwin Whatley, Linda Polgar, Eleanor Powers, Laura Besvinick, Scott Atherton, Joseph Levine, Gina Maranto, Arun Swamy, Paul Levenson, Dan and Helen Levenson, and my siblings, Richard Levenson, Irene Levenson Girton, and Leo Levenson.

Everyone mentioned, and many more besides, made the experience of writing *Ice Time* enjoyable, as well as possible—and to them go all my gratitude and the credit that is due. Any errors of ommission, commission, emphasis, or interpretation, must, alas, rest with me.

TOM LEVENSON

Preface

This is a book about climate, the origins of the earth, and its development into a planet that nurtures us; it is a book about the attempt to understand how that planet came to be, how it functions now, and what may happen to it; it is a book about the act of trying to make sense of the world, the demands we make of knowledge, and the people who seek it.

A storm is blowing through Boston as I write this, and a little snow is falling now. This January has been one of the snowiest months in recent memory, and there is more to come, the forecasters say, in February. The weather is something we can do nothing about, other than endure it or enjoy it, but it is always accessible as human experience. It happens to us all the time, through fluctuations of temperature, in rain or snowfall, under cloud cover, or in the sudden expressions of fury that mark the great storms. Weather is clearly something that takes place in the atmosphere—winds blow, carrying good weather or bad to where we are, fixed on the surface of the earth. The events that make the weather then reach down to touch us. Above all, though the weather fits into certain patterns (summer is warmer than spring, San Francisco far foggier than Boston), our day-to-day experience of it is one of chance. Tomorrow might be warmer than today, but of that we cannot be sure for reasons intrinsic to the way the atmosphere behaves.

But weather emerges out of events that take place in much more distant points in time and space than does the current weather, today's snow, the chance of tomorrow's thaw. Over time weather becomes something other than the feel of the wind against our cheeks as we step outside. As it encompasses more of the globe, more of time, weather becomes climate, the broad conditions that year in and year out define what we may expect the outer bounds of the behavior of the atmosphere to be over a particular piece of land—the climate in San Francisco that is clearly distinct from that, for example, of the American Northeast. At least that is the human conception of climate: the sum of the accumulating days of weather.

There is, however, a new view of climate and of climate's relationship to weather, and it is with that new understanding that this book is primarily concerned. It is a view that looks at climate on the very longest time scales and then focuses in closer and closer until it becomes clear how the apparent accidents of the weather are connected to the evolution and continued functioning of a climate machine. And that machine—or, better, system—is not one confined to the atmosphere; rather, it encompasses land and sea and the living things that inhabit them and even reaches into the interior of the earth, until ultimately it includes the planet as a whole.

This view of climate is still in the making; its creation hinges on a trick of perspective. As long as the question that interested a scientist or a whole science was something on the order of what makes a thunderstorm, climate in this novel sense remained inaccessible. Such a question allows the researcher to use the traditional methods of experimentation in which everything but the specific phenomenon under study is held constant to explore how that phenomenon, the thunderstorm, actually works. In this case, meteorology is more of an observational than an experimental discipline, but the principle remains the same—look for one set of events that explain one phenomenon.

The new inquiry introduces a different perspective: Climate has become the unifying link for the sciences that examine how the earth works as a planet, how it functions as an entire system through every portion of which energy and matter circulate. The conditions manifest themselves as weather. The dynamics of the atmosphere —created by the interaction between the surface of the earth and the energy from the sun, by the flow of water from the oceans into the atmosphere and onto land, and the sudden eruption of storms of intense energy—are all bound within a system of climate that has been and is still emerging over the course of the entire history of the earth.

It is climate that has created the conditions within which the human species has come to thrive, yet this idea is resistant to the classical form of scientific investigation. A single experiment cannot add much to the understanding of how climate, weather, and life interact because it is impossible within a web of such complex connections to hold all but one process constant and observe the result. The science of climate that has sprung up within the last generation has had to create its own questions, its own novel form of experiment, and, above all, its own strikingly new vision of the planet seen first whole and only then split into its component parts, its thunderstorms and fog and sunny days.

With this perspective it becomes possible to understand the world and our place within it in a far richer way than we did even twenty years ago. The proximate cause of the revolution in our conception of climate has been the development of a set of tools—computers and satellites and numerical methods to make sense of data being collected more quickly and thoroughly than ever before. These tools analyze motion, chemical changes, the physics of storms and seasons, and the variability that characterizes climate on every time scale, from minutes to billions of years.

This book opens, then, with a consideration of what the crucial, driving elements of the climate system are, how they evolved, how they work, and how, ultimately, they produce both the weather and the larger context of an environment that allows us to prosper. Part I begins, appropriately, at the beginning, with the development of the atmosphere into a place at least vaguely recognizable to us. From there it follows the emergence of ancient climates and their transformation into the climates of human memory and finally into the weather that is our daily prospect.

The development of this knowledge of climate comes at a crucial time within the history of the earth. Human societies can now alter the weather and for the first time stand in real danger of altering climate in the larger sense as well. Consequently, we will change the kinds of lives it will be possible to lead on earth. The emerging discipline of climate studies provides a lens through which to examine our actions. Part II will explore the revolution in method that has created that lens; Part III will examine what that lens reveals.

Taken as a whole, this book is about science first. But there is more to science than particular pieces of knowledge that the exercise of skill and reason has accumulated over time. Research into this area yields a return far greater than improved weather prediction, for with this new science we see not just more detail but an entirely different world, a place transformed by the very act of seeing. The earth looks different, for example, when we learn that events in the south Pacific in one season will dictate whether Louisiana floods half a year later; different compared to a world in which you imagined rain falling was just an accident of time and place. And with the change in what we see, we change who we are, alter our relationship each to each other. Climate and our knowledge of it define our planet, our home. As we reshape our climate with our daily acts, how we think of it reveals, finally, what and how we value ourselves.

PART I

ICE TIME

CHAPTER 1

In the Beginning

IN THE BEGINNING was violence, change, and possibility. There was a star, with a ring of dust. The dust motes collided with each other, coalescing into pebbles; the pebbles became rocks, the rocks, in orbit around their star, grew into planetisimals. In the beginning, about 4.7 billion years ago, those planetisimals became planets: the third one from the star was ours, the earth.

Rock is the ultimate historian—what it is, and what remnants it contains are the only records of what the earth was like through virtually its entire lifetime. But almost no rock remains in the earth's crust dating from the first billion years of the planet's existence; in fact, almost nothing tangible remains from that time. As a result, for those first billion years one must simply reason backward. The image of the world then becomes, in large part, a work of the imagination.

Billions of years; tens, hundreds of millions; thousands of years; a century; a season. The earth today—and the earth when Tyrannosaurus rex chased down its prey—is the product of a weave of time. We experience climate as weather, the day-to-day accumulation of high and low temperatures, winds, clouds, rain, and sunshine. But what we experience day to day emerges from a context that becomes recognizable only as our perspective widens to encompass greater and greater stretches of the history of the earth. Climate is a matter of scale, or better, of the intersection of events that occur on any number of different scales of time and distance. A system of climate and life based on oxygen takes aeons to establish; the broadest conditions of warmth or cold are set over many millions of years; ice ages come and go almost skittishly quickly by geological reckoning; and today it's bitterly cold, New England cold, as I write this, but it might warm up just a bit tomorrow. All of the different processes are at play, overlapping to produce a unique catalogue of specific events: a Boston winter day.

In science, rather than simply out there in nature, this winter day, seen as a window on all the history of the earth, poses a novel and extremely difficult problem. The science of weather is an old one, one of the oldest humans have tried to study, for obvious and practical reasons—it is, after all, a great help for a farmer to know how tomorrow will differ from today. But weather prediction has never been a perfect science, and it isn't one now. We must turn to a different line of research to discover what we can and cannot learn about the weather, and the roots from which the weather springs.

The science of climate is new and has begun to change not only the definition of what climate is but our own image of the world. The older definition, the commonsense understanding of climate, was a static one: climate is the broad array of atmospheric conditions that obtain in a given place at a given time. New England has harsh winters and warm summers—that is its climate. The United States lies in a temperate zone, and Costa Rica is in the tropics: their basic conditions do not vary. This definition of climate used to be scientific orthodoxy; it took Louis Agassiz, for example, several years before his idea that Europe had once been covered with ice could overcome the skepticism of his learned peers, who had experienced nothing but the relatively mild weather of Europe in the mid-nineteenth century. Climate as a science, a distinct discipline, did not exist. Meteorology, atmospheric physics, oceanography, and geology all glanced off the problem, but the study of long-term changes in the climate tended to remain purely observational, just the leisurely accumulation of facts in one field or another. The idea that a science of climate could link all these inquiries together remained, until very recently, just out of sight.

The new view, developed in its richness only in the last decade or two, is of climate as a dynamic and inclusive system. Climate is the sum of all the processes that integrate the disparate givens of the world—soil and rock, oceans, the atmosphere, life—on all scales of time and space. Events in Peru reverberate in San Francisco; events that began five minutes ago have their roots in processes that can be traced back as far as evidence exists on earth. In its very broadest definition, the study of climate is the study of how the land and ocean and atmosphere together change over time—over minutes (when the weather shifts) and over billions of years (as climate for the earth as a whole takes shape and becomes familiar).

At the very beginning, this longest scale is the most significant, partly because no finer level of detail shows up, but more because it is on this billion-year schedule that the basic parameters within which the planet evolved were laid down. Climate existed, after a fashion, from the very earliest days on earth, but any reconstructions of it depend on arguments by analogy. The original atmospheres of all the planets would have contained essentially whatever gases were in the neighborhood as the solar system formed. Jupiter today is large enough to have hung onto much of that primeval atmosphere; in particular its gravitational pull is so strong that it is able to retain an atmosphere rich in hydrogen, an element so light that its atoms eventually simply float away from smaller planets. The early earth could have, the story goes, resembled a tiny Jupiter, with a gassy envelope rich in hydrogen and helium along with such currently rare elements as argon, neon, and xenon. The comparatively weak gravitational pull of the earth allowed hydrogen to escape out to space then, as it continues to do today. The modern atmosphere also loses gases downward when, for example, atoms of oxygen bind with atoms of iron to form ferrous oxide (rust); similarly, the ancient atmosphere must have lost some of its gases to chemical reactions with elements in soil and rock.

Even if that first atmosphere vanished altogether, though, the earth would not have remained naked for very long. Volcanoes would have been erupting, and each eruption would have produced gases and steam. Over time, the volcanoes would have replaced whatever remained of the primeval atmosphere with one made up of water vapor, carbon dioxide, nitrogen, and hydrogen, bound up with sulphur and other minerals. Such an atmosphere may have made for a fairly warm earth. Although the sun was less intense as a young star than it is now—it was possibly as much as a third cooler then—the geological record indicates that liquid water existed on earth back to the very earliest time for which rocks exist to supply clues. That means that the earth had to be within 15°C of the temperature it is today—neither like Venus, so hot as to boil all water away; nor like Mars, cold enough to keep the thin polar ice caps frozen solid. To maintain a relatively warm climate, given a weak sun, the earth's atmosphere had to trap and hold what heat was available considerably more efficiently than the modern atmosphere can.

The abundance of carbon dioxide in the fumes of volcanic eruptions provides the key. Carbon dioxide is a greenhouse gas: It lets in sunlight, but when the earth remits infrared radiation (heat rays) toward space, carbon dioxide molecules in the atmosphere absorb the infrared wavelengths relatively efficiently, trapping the warmth. The early atmosphere was rich in carbon dioxide, and, at least for a time, the levels of the gas would have continued to build up unchecked in the atmosphere with each eruption.

It is, once more, almost impossible to say just how warm this original greenhouse effect would have made the earth. The water vapor that accompanies carbon dioxide in volcanic emissions could have made the early earth a fairly cloudy place, in theory, at least; if they did exist, those clouds would have cast shadows that would have blocked some sunlight out, thus cooling the earth. There are, however, wheels within wheels. The clouds produced rain, and at some point in that first billion years, the accumulated showers began to create the first oceans. Oceans are relatively dark and so reflect less of the sun's energy back into space than does dry land, so the clouds that could keep the earth cool would produce oceans that help warm it up; the oceans then may provide a source of more water that could evaporate, form into clouds, and begin raining all over again—and so on, without ceasing, in a dance that continues in a radically different world today.

The arguments go on, planet-sized "what if" games, about how cloudy the earth actually was, how much carbon dioxide appeared how quickly, whether other gases might have formed greenhouses of their own. With only the barest clues remaining, these questions have no clear answers. It comes back to this: Early in its history the world was warm enough, at least in places, to permit the existence of liquid water.

For liquid water read life. The precise details of the origins of life remain hidden, but at a minimum life requires water, and if it was warm enough to permit the existence of ponds or seas, then the conditions in which living things most plausibly first emerged have existed as far back as the rocks can reveal. And for life read, in embryonic form at least, climate, for, as independent of biology as it seems, the climate system we recognize requires the existence of life. From the beginning, the creation of the atmosphere, of the air we find fit to breathe, hinged on life.

Since it first appeared in the fossil record, life has persisted on earth without a break. That fixes absolutely, in the absence of any other evidence, some crucial facts about the history of the earth's climate. Over the three-and-a-half billion years in which life has existed, our planet's climate has never varied enough to wipe out everything. Fairly liberal boundaries are involved here. Broadly, life could persist within a temperature band running from 0° to 40°C, and some forms of life can withstand temperatures as low as −60°C or as high as 60°C. Still, the persistence of a broad spectrum of living things means that while the world has changed, and at one crucial point was altered abruptly (at least it seems abrupt when our perspective is aeon by aeon), in the broadest sense it has remained a consistent haven.

Species have come and gone in dizzying numbers, most vanishing probably without even leaving fossil traces to mark their presence and their passing. That is the essence of life: flexibility in the face of change, a malleability that seems foreign to rock or a mantle of gas. It is clear enough, at first reading, that living things bend to their environment, not the other way round. Yet the crucial intellectual leap that has created climate science is the recognition that an environment and the inhabitants of an environment are bound together: life does not merely adapt to inanimate surroundings; in a kind of geological dialectic, living things bend the conditions in which they live as they bend to them.

That idea is indisputable now, on a certain scale. A human being warm and dry within a building during a Boston snowstorm has modified his environment in such a way as to create a (micro) climate that suits him better than the fare served up outside. The inspiration for the new science of climate, however, was the recognition that some kind of environmental engineering is intrinsic to all life at all times; in fact, the interaction of life and nonlife creates climate. From the beginning that interplay molded the atmosphere into one that, turn and turn about being fair play, helped trigger a transformation in the nature of life on earth.

Almost certainly, there was no oxygen free in the atmosphere at the very beginning of the earth's existence. No oxygen means no ozone, a gas whose molecules are made of three oxygen atoms bound together, rather than the usual two. Ozone absorbs ultraviolet (UV) light. Thus, in the earth's early days, UV radiation from the sun was free to strike the water molecules in the atmosphere instead of reacting with ozone, as it does today. UV light splits water into its components—two atoms of hydrogen and one of oxygen. Most of the resulting free oxygen atoms must have recombined into ordinary oxygen gas (O_2), but some portion of them formed into the three-atom molecules of ozone, which began the building of a shield that could filter out the extreme doses of UV radiation before it reached the lower atmosphere and the surface of the earth.

For perhaps as long as a billion years, before life existed, this game of billiards between atoms and light was the only source of free oxygen in the atmosphere. At best, it could have produced only traces of the gas, so that when life first emerged, it was probably in a warm, wet world, possessed of an unbreathable (to us) cloak of an atmosphere. It was a world that we can reconstruct, broadly, because, in the absence of life, physics and chemistry (and the examples of neighbors like Mars and Venus) provide a set of rules within which any combination of elements would have had to play.

With the emergence of life those rules change and the game becomes vastly more complex. On earth the presence of life for so much of the planet's history has obliterated the memory—the record written on the planet's surface—of what a world without life looks like. It is hard to see what life does within the climate system while we stand within that system. It took an analogy and a shift in our view of the earth to begin to draw the first major pictures of how climate and life function in tight embrace.

In the early 1960s, an English chemist named James Lovelock collaborated with scientists at NASA's Jet Propulsion Laboratory to design experiments to be carried by the Viking landers to Mars that could determine whether life exists on the fourth planet. This led Lovelock to the question of how one would detect the presence of life on earth if one were to come upon us unawares.

Suppose you were some alien scientist designing a probe to explore the earth, a planet you could not see, about whose chemistry and evolution you knew nothing. How would you go about producing a set of experiments that would confirm, no matter what, that life exists on that planet? If you can be sure that your probe would be able to recognize New York City and be able to put down at Times Square at noon, then your design problems would be easy—just put a set of cameras in and beam the pictures back, and you'd probably feel confident that you could pick up incontrovertible signs of some kind of life-forms. But the probe might land in the sea or on the Gobi Desert—or it could, as a worst case, land downtown in my home city of Boston after the businessmen's bars have closed, at which time there would be no signs of life till the next morning.

These kinds of difficulties rapidly led Lovelock to conclude that sending a probe to Mars in the hope that a little green man would appear on cue was probably the wrong approach. And so, turning the question back on itself, he asked if there is something that living things do that leaves traces of their presence that are more easily detectable than direct physical evidence. He reasoned first that life can be defined most broadly as a member of a class of phenomena that can use the energy available in its environment and must, as a result, excrete the end products of that energy use. All such animate machines are chemical processing plants (in fact, living matter on earth processes one-third of the total number of naturally occurring elements found on the periodic table), and as such they modify their surroundings by their patterns of consumption and waste production. Thus it ought to be possible, Lovelock concluded, to detect life by looking for certain chemical compounds—the by-products of living metabolisms—in the atmosphere of the earth or Mars (soil or water would work, too, but the atmosphere is the most accessible place on another planet) that shouldn't be there if the only source of chemical reactions and change was the billiard game of physics and inorganic chemistry.

Then Lovelock noticed that the earth's atmosphere today is chemically truly remarkable in that it possesses abundant oxygen and methane, two of the most significant metabolic wastes of earth's complement of living things. Since the two compounds would normally react with each other to form carbon dioxide and water vapor, the presence of the two gases in their current amounts is, according to Lovelock, almost impossible on a planet without life. Life had to be producing quantities of both gases greater than that which can comfortably react within the atmosphere. "The only feasible explanation," Lovelock wrote, was that the earth's atmosphere "was being manipulated on a day to day basis from the surface and that the manipulator was life itself." Lovelock subsequently teamed up with the pioneering evolutionary biologist Lynn Margulis, and between them they recast an older conception of the earth's history. They claimed that there exists "a complex entity involving the earth's biosphere, atmosphere, oceans, and soil; the totality constituting a feedback or cybernetic system which seeks an optimal physical and chemical environment for life on this planet." In other words, life and the rest of the physical planet together form a kind of superorganism that in its entirety maintains the conditions that best suit life on earth. Following a suggestion from the novelist William Golding, Lovelock gave his planetwide entity a mythic role, as a preserver of a haven for life on earth, naming it after Gaia, who in Greek mythology was the original earth-mother whose body nurtures us all.

Lovelock and Margulis are at pains to call their idea the Gaia *hypothesis*, not theory; it is a suggestion to be tested or, at its furthest limits, a mirror for some image of a world-scale system that encompasses life and nonlife. Climate, in this scheme, is produced and maintained by the continual interplay between the living and their environment. Even the simplest life-forms, the very oldest forms known, they note, have played leading roles in the creation of the earth's climates and have, in fact, left their chemical signature scrawled in rock across time counted in billions of years.

Of all such chemical signatures, the most evident today is, as Lovelock first realized, the entire atmosphere, containing its remarkable proportion of oxygen. Oxygen now accounts for about 20 percent of the total volume of the atmosphere, and virtually all of it is produced by living organisms. Yet to early organisms, oxygen was a poison, and modern plants and animals have had to evolve a variety of sophisticated shields and filters to guard against its power. Without them, an oxygen-fueled metabolism would lead to the buildup of such compounds as hydrogen peroxide, which kills unprotected cells.

The very earliest organisms and the simplest single-celled ancestors of modern bacteria neither used nor produced oxygen; they derived the energy they needed to live from fermentation, as yeast does when it breaks down the sugar in grape juice and leaves behind wine. All more complex forms of life use oxygen in a complicated series of chemical steps that transform sugar into a form of energy living cells can use, a process called respiration. Respiration appeared on the scene when at least one form of life that used fermentation, the so-called blue-green algae or cyanobacteria, gained the ability to produce and exploit oxygen.

Blue-green algae existed at least 2 billion years ago. Like modern plants, they were photosynthesizers; that is, they were able to take energy from the sun and use it to power the production of the chemical compounds they needed. Photosynthesis produces sugar out of carbon dioxide and water, with oxygen gas released as a kind of exhaust fume. At first the oxygen from the algae would have dissolved in the oceans, but over time it began to leak out into the atmosphere. Blue-green algae cannot tolerate much oxygen (mats of the organism exist today only in areas of the ocean where the amount of oxygen remains low), but they offered a novel and potentially extraordinarily desirable technique of exploiting the energy available to any organism that could harness it. Oxygen-based metabolisms were now an option.

Thus began a fitful buildup of oxygen. Geologists have found spread through Africa and the Americas deposits of a mineral called uraninite that lie in river sands more than 2.3 billion years old. This mineral binds easily with oxygen; the fact that it was common in an unoxidized form then implies that 1.5 billion years of life had not yet added much oxygen to the atmosphere. However, about 2 billion years ago the first "red beds," made of sands that gained their color from iron that had rusted in the presence of oxygen, appeared.

There is no reconstructing such a world in detail; any picture of it is drawn largely from the imagination. But algal mats remain today in a few places, and the images they inspire may not be too far off. Think of a shallow lagoon filled with strange structures—mats floating, with perhaps a trunk of matted material reaching down below the surface. Although we have no good record of the temperature through most of these first 2 to 3 billion years, it may have been very warm, so imagine a tropical feel to this lagoon. The weather is humid and frequently rainy, and storms across the seas may ruffle through the masses of algae and may even help spread the mats about. It is monotonous: there is water, in the ocean and falling from the sky; there are the algae; and there is, undetectable simply by watching this small lagoon, a fitful, ongoing change in the air that blows across the wave tops and the mats. The algae are everywhere, and everywhere they are creating a world for which they are remarkably ill suited.

With the spread of the blue-green algae, oxygen in the atmosphere probably reached a level of about 1 percent of its content, compared with just over 20 percent today. That's enough to produce the red bed rust and enough to begin establishing an ozone layer that could block ultraviolet radiation. Over the next half a billion years or so, the accumulation of oxygen proceeded fairly slowly. Blue-green algae were still its main producers, though as organisms betwixt and between—able to survive either with or without an oxygen-based diet—they were not terribly efficient at pumping the gas into the environment. However, around 1.4 billion years ago, the game changed once more: A radically new form of life appeared on the planet, and as it prospered the pace of change in the environment picked up as well.

The newcomer was the eukaryotic cell. Bacteria and simple algae are organisms whose cells lack a nucleus. Everything they are—their metabolism, their genetic material, their bodies—is all crammed into a single cell called a prokaryotic cell. The eukaryotic cell, in contrast, contains within it several smaller, walled-off systems that perform many of the functions that keep it alive. Most important, a eukaryotic cell contains a nucleus, which houses the genetic material of the organism, packaged into a series of chromosomes.

The eukaryotic cell marks the single greatest leap in evolutionary history because the distinction between it and prokaryotic cells is greater than that between any of the myriad species that have since emerged. Certainly, the eukaryotic cell made the world a much more interesting place in which to live. Before its appearance, life on earth had a certain monotony to it; some variation had developed in prokaryotic cells, but nothing to match the explosion of diversity that the eukaryotic cell brought to life. Most important, eukaryotic cells are easy targets for the variation that comes from sexual reproduction. With their genes packed neatly onto chromosomes within a distinct subunit, the nucleus, they made sex a broadly accessible reproductive technique. (Bacteria can also have sex, but prokaryotic reproduction involves, for the most part, cloning identical daughter cells from the mother cell.) From the first eukaryotic cell descended a host of other single-celled organisms, together with all multicellular beings, every plant and every animal.

It has been a long road home, but we can finally return to the problem of why the air is fit to breathe. Eukaryotic cells accelerated the biochemical cycle that leaked oxygen into both the oceans and the atmosphere, that had been established under the old prokaryotic order. In what may have been a feedback loop, as oxygen built up, so did the pace of biological change: As more of the rust-colored red sands appear as geological strata, more and more fossils of novel life forms remain in the rocks. It looks, writes biologist Preston Cloud, "as if evolution were speeding up."

Absolutely. Blue-green algae and their ilk ruled the world for perhaps as long as 2 billion years; eukaryotes then took over, but the first single-celled eukaryotes dominate the fossil record for no more than 500 million years or so. At that point, about 700 million years ago, the fossils of multicellular animals begin to appear. By extrapolating from assumptions about their metabolic requirements, it looks as if the oxygen content of the atmosphere had to be about one-third of what it is today. The pace of change continued to speed up over the next 300 million years, when land plants appear for the first time along with large marine fish. At that point the oxygen content of the atmosphere appears to have reached the current level, about 21 percent of the total volume of the atmosphere.

And it is a wonderful world, for us at least. But the evolving atmosphere has emerged only at a price: Blue-green algae and all the other creatures that ruled the earth for the longest uninterrupted stretches of time are now extinct or confined to tiny exceptional ecosystems. But if in Lovelock's hypothesis Gaia maintains the earth as Goldilock's paradise—not too warm, not too cold; not too hard, not too soft—providing constantly the "optimal" conditions for life, we must now ask, optimal for whom, or what? If

Gaia exists, the transition from an oxygen-poor to an oxygen-rich ecosystem implies that the entity made a choice (and hard luck for blue-green algae). Anthropocentrically, I can only applaud the decision, if such it was; I enjoy the variety of life that the eukaryotic cell and oxygen together make possible. But at the same time it is difficult to conceive who or what could have made such a choice for the entity as a whole: not the algae or the rocks or the oceans or the gases that surrounded them.

Nevertheless, oxygen, once present in the atmosphere, does become one of the prime pieces of evidence that Gaia, or something like it, does exist. In the billion years between the emergence of the first complicated cells and the creation of an atmosphere that we could breathe today, life and climate established a feedback mechanism. Change in the chemistry of the atmosphere could only go so far before life—at this juncture mostly plants and bacteria in the ocean—responded by bringing the climate system back to a rough state of equilibrium.

The mechanism works like this: Oxygen enters the atmosphere when a plant uses the carbon from a molecule of carbon dioxide, and releases the oxygen. However, a plant only stores carbon temporarily, at best until it dies. Then, as the organism decays, the carbon atoms that had formed part of the body of the plant become free again to bind with oxygen. In theory every oxygen molecule that escapes from a living organism should be bound up again with an atom of carbon from a dead one, leaving little or none left over for the atmosphere.

Nonetheless, the air we breathe remains oxygen-rich. It does so because there is a break in the cycle that allows a certain amount of oxygen produced by photosynthesis to remain behind in the air. Some carbon in the ocean never encounters a molecule of dissolved oxygen as it sinks from the surface down into deep water. It gets buried in the accumulating sediment and eventually locked up in newly formed rock. Each atom of carbon that ends up as part of the seafloor translates into a net gain of one molecule of oxygen gas for the atmosphere.

This cycle allows for remarkably precise fine-tuning. When oxygen levels drop in the atmosphere, less oxygen reaches the ocean. This means that even less of the carbon fixed in ocean plants would be oxidized in the decay process as the plants' remains sink.

In turn, more carbon gets buried unoxidized, which means that some of the oxygen produced by living plants is free to leak into the atmosphere. Similarly, when oxygen levels swing upward in the atmosphere, more atmospheric oxygen dissolves in seawater. Then finally all or most of the carbon eventually binds with oxygen, which makes room for more oxygen from the atmosphere to dissolve into seawater, which eventually brings atmospheric oxygen back to an equilibrium.

And round and round the cycle goes. The balance between the oxygen produced by living beings and that consumed in the decay of the dead maintains atmospheric oxygen levels right around the 21 percent mark. There is an almost eerie precision here, because that amount of oxygen is neither too much nor too little, but just right for modern forms of life. If there were much more, oxygen might feed fires so intensely that they could burn without ceasing; a single lightning strike might be devastating. If there were much less, large, complicated forms of life asphyxiate; we would be left with the simpler, ancient organisms that can do with less or nothing. We could not have done better for ourselves—we meaning the whole variety of modern forms of life—if we had planned it that way.

That sounds like Gaia incarnate, a proof for the hypothesis; actually, it is Gaia drastically exaggerated. And yet it is perilously easy to be convinced by an image of progress even though that image turns out to be a mirage. Just because one event succeeds another—an oxygen buildup first, the birth of Descartes second—it doesn't follow that the first happened as part of a design intended to produce the second. "I breathe, therefore I am" might not raise eyebrows philosophically, but as a foundation for climate history such a statement is nonsense.

That's the heart of the criticism that other scientists level at Margulis and Lovelock. The broadest statement of the Gaia hypothesis—that some system of feedback mechanisms exists that by its design automatically seeks out the optimal conditions for life—begs the counterargument. Optimal conditions are always in the eye of the beholder. Plenty of species have disappeared—most of the species that ever evolved, in fact. Life has plenty of losers for whom some combination of conditions, changes in the weather, alterations in plant cover, variation in the competition from other species have added up to a situation far from perfect, or even adequate.

And yet we are here, and with us a fantastic diversity of life, all thriving in a climate system that has maintained itself within rough boundaries for hundreds of millions of years. If it is a mistake to assume that oxygen appeared specifically for our benefit, it is also difficult to escape the conclusion that this careful regulation of the climate seems to have some kind of purposefulness about it. Stephen Schneider, a climate theorist who is a friendly goad to both Lovelock and Margulis, offers Gaia a sideways compliment: "As religion," he writes, "I find Gaia deep, beautiful and fascinating."

As religion! That's one of the nastier things one scientist can say about another's work. But there is more to Gaia than religion, however much the name itself invites comparison with and exudes the whiff of ancient belief. The specific idea that some single superentity organizes the living world as well as the air, soil, and water that envelop it, into a self-regulating system has its problems. Yet the approach, the core idea, remains suggestive. That central thought is deceptively simple—the recognition that feedback systems exist between life and its environment that affect the future of both.

Such a notion seems a tiny revelation at best, too simple, and too obvious. In fact, Lovelock's is a bold cry, stated almost overboldly. Lovelock noticed that a whole planet, over millions and billions of years, can behave in a way that makes it at least appear to be a living being, whether or not closer examination uncovers a single, identifiable earth-mother. Most people never conceived of the world in this way and simply could not take the necessary leap of imagination to go from questions of atmospheric chemistry, for instance, or the geology of ancient sediments to bind each with the other into a coherent whole. There is a direct analogy here to the old belief in a flat earth. From ground level the world does look pretty flat; it took a leap of the imagination to step out of that frame of reference to imagine the world as a ball.

The Gaia hypothesis is a symptom of the sea change that has overtaken the field of science concerned with the creation and maintenance of the conditions that we feel as weather (a storm today, fine sunshine tomorrow) and that we call, as they average out over time, climate. What happens over billions of years creates the air I breathe as I write these words. With this shift in vision has come a view of the earth made animate. Climate as a concept now encompasses the whole world, life and nonlife, altering itself as each of its parts moves and changes, forces events and bends to them.

In "Intellect," Ralph Waldo Emerson wrote that "Nature shows all things formed and bound. The intellect pierces the form, overleaps the wall, detects intrinsic likeness between remote things, and reduces all things into a few principles." Emerson, or Thoreau, or Muir recognized the existence of connections, of systems in which the chance occurrences of a life on a single day are embedded. But they and their contemporaries could do no more than recognize this fact and express their faith that the mystery of connection would eventually yield to investigation. The new science of climate is the product of half a dozen sciences. For instance, the story of oxygen hinges on new findings in geology, on the seminal biological work that traced the evolution of oxygen-metabolizing organisms, and on the race in space with its accompanying pressure to think about what distinguishes the earth from any other planet. In sum, the emergence of climate science is the product of a evolution in science. Before the 1960s, even if one thought about the world in this way, it was impossible to piece together the puzzle.

As for Gaia itself, it should come as no surprise that the idea emerged only when it did. There are very few memories that everyone shares. For my parents' generation it is V-E Day; for those somewhat younger it is the day that Kennedy was shot. I am too young to recall that event, but I remember precisely where I was and what I was doing on the day the words "The Eagle has landed" were relayed. And I remember with absolute clarity the first sight of the earthrise over the moon. At that moment, seeing the living planet shining over the dead rockscape, the idea of Gaia became simple and obvious, even if the details of the chemistry and biology remained to be worked out.

It is those details that make the idea live, of course. They have generated the original conception of the Gaia hypothesis, together with many other theories, including one from Stephen Schneider called coevolution. According to Schneider, life and environments evolve their feedback systems in conjunction with each other, but without the sense of purpose or direction implied in Lovelock's original hypothesis. These competing ideas share the theme, though, that their proper study is the whole planet, and they gain their indisputable power not just from their details, but from that picture of the earth, whole, blue, and unmistakably alive—a picture that carries the emotional force of the idea and of the science that attempts to explain it.

The history of the early earth is impossibly remote, a few tag lines written in chemical hieroglyphics across a scattering of ancient rocks. The life and death of blue-green algae do not affect in the slightest, at first reading, the price we pay for a dozen eggs. Gaia as a name belongs to myth; as science it thrives as metaphor, as a spur to see the world whole.

CHAPTER 2

Of Air and Stone

THERE WAS ICE on earth 2.3 billion years ago, vast sheets of ice, glaciers, an ice age. There were ice ages again running through a period of 300 million years beginning about 915 million years ago. The glaciers came again at least twice over the next quarter of a billion years. And then the earth grew warm; throughout the reign of the dinosaurs the world was apparently free of large, permanent oceans of ice.

There is ice today, great expanses of ice covering Greenland and Antarctica, floating masses extending outward from the polar seas, little glaciers creeping down Mount Rainier, the Alps, and the Andes, even Kilimanjaro. The last ice age ended perhaps 12,000 years ago, but it was one of a series of regular occurrences, following cycles that persist to the present day, each of which runs 100,000 years or so. The earth is much cooler today than it was tens of millions of years ago, and even at the warmest ends of this cycle, continent-sized glaciers remain.

Climate on every time scale reflects a basic struggle between constancy and change. On the longest time scales, over billions of years, the balance leans toward constancy; the overwhelming fact is that the world has remained enough the same for life to persist. Shift the scale, alter the perspective to tens of millions of years, and the inquiry shifts as well. Seen at (somewhat) closer range, more detail emerges, and with that detail comes a transformation in the basic problem of climate science. Where the issue once had been how conditions could remain broadly the same for so long, it now becomes a question of why, and how, within those broadest bounds, does climate change? How warm or cold can the planet get? What drives change within the system, and what restricts the degree of change possible at any time? In practical terms the goal here is to explain specifically the changes in global average temperatures over long periods of time. On this scale other climate issues—how wet or dry a region was, for example—don't show up as readily in the ancient geological record as the great ice ages do. The problem requires an attack on two fronts. First, one must reconstruct the climate history of the planet on a multimillion-year time sequence, which is difficult enough; then, one must try to assign causes for events whose mere outlines are almost impossibly remote and difficult to discern.

Within these particular questions raised by the temperature history of the earth lies an even more basic issue: How predictable is nature? Clearly, on some time and geographical scales, events are purely random. If I were to walk outside right now, I would have absolutely no clue, not a chance of guessing precisely when I would feel the first puff of a breeze against my face, nor how strong that breeze would be. At the same time, because the atmosphere is a reasonably well-mixed medium and the oxygen content of the atmosphere is controlled by a process that extends to the roots of time on earth, I have no problem believing, with absolute confidence, that for the rest of my life I will breathe in virtually the same amount of oxygen with each breath no matter where I travel (as long as I stay at sea level, of course). Between the billion-year time scale and that of a few seconds the earth can be varyingly capricious. Certainly climate changes on all time scales. The study of those changes is in fact a quest for the mechanisms that on each time scale may regulate the pace and nature of the change, a search for the particular physical explanations that must underlie any attempt to predict the behavior of nature.

For the great ice ages the first approximation of an explanation depends on which rocks you ask. The last ice age has left its telltales written quite clearly across the landscape. When Louis Agassiz first promulgated his heresy that ice had once covered the Swiss countryside, he looked to the valleys there that retain glaciers to this day. Like other observers, he noted the presence of strange boulders, called "erratics," tossed down in valleys like flotsam after a flood (for some, Noah's flood, *the* flood) had drained away. He saw the strange polish along the bedrock—a sheen imparted as if by some massive swipe of sandpaper; he saw the debris of rocks and boulders fringing the margin of existing glaciers. He saw what can be seen still, markings in stone that indicated that ice once flowed over vast stretches of land now clear and verdant.

The first great glaciations must have scored the earth as deeply in their turn, and, in principle, we ought to be able to track the history of the early ice ages by following the same reasoning Agassiz used to persuade himself and his contemporaries that ice once covered Europe. But the marks left by these earlier glaciations are quite subtle, tracks turned ghostly with great age. There are, however, telltale deposits of ancient rocks that strongly suggest that they had been ground together and laid down by the spread of ice. Such rocks have been found in North America, Africa, and Australia, and their ages seem to hover around the 2.3 billion-year-old mark. That date and their spread are vague enough, however, to make it almost impossible to determine just how much of the earth was icebound during the possible range of time in which each of the glacial deposits was formed. Uncertainties about both the timing and the extent of these glaciers also muddy the search for the cause of the ancient ice ages. The record is so spotty that geologists are not sure whether areas near the equator or places nearer the poles were the coolest places on earth. It's also possible that volcanic eruptions had tossed enough dust into the atmosphere to screen out sunlight and cool the earth. While some of the glacial records in the rocks do indeed contain evidence of volcanic activity prior to the buildup of the glacial debris, others do not.

Such traces are the currency of science—data—and like money, richness of data both buys you some credibility and ties you down, eliminating at least some theoretically plausible explanations. For this early period, so little data exists that it is hard to prove anything, so theorists have come up with a variety of ideas to explain the ancient ice ages, all elegant and mostly immune to both proof and criticism. For example, a change in the earth's orbit could have reduced the amount of sunlight reaching the planet. However, the only physical signature of such an event that would show up in the rocks would be the marks of the glaciers themselves.

That brings us full circle: It's the presence of the glaciers, in this scheme, that explains what caused the glaciers to appear. The Australian climate historian L. A. Frakes has prospected through the various theories proposed to account for these early ice ages. He isn't terribly enthusiastic about any of the possible culprits, but his choice for the least unlikely of them all emerges out of the recent revival of what was once a radically unorthodox idea, that continents drift over the face of the planet. Frakes argues that the glaciers originated at sites near the poles and that the ice ages began because the continents of the early earth had drifted to positions that took more and more of their land nearer to the polar regions. More land near the poles meant that more precipitation fell as snow and could be compacted on land to form glaciers. With enough glaciers, the increase in the amount of sunlight reflected back into space off the glistening white sheen of the ice effectively reduced the amount by which the sun warmed the earth, creating the feedback loop by which the growth of glaciers encouraged the growth of more glaciers.

Whatever the precise cause, the cool period that began around 2.3 billion years ago broke after perhaps 100 to 200 million years. If Frakes is right, then the reason for the end to this primordial ice age could have been more continental drift, this time carrying landmasses away from the poles toward warmer regions where glaciers could begin to shrink, thus interrupting the feedback cycle. For the next billion years or so, the rocks preserve a temperature gauge that indicates that the earth on average was significantly warmer than it is today. The mixed clumps of rock that would indicate the passage of a glacier are missing, and the scratches on exposed bedrock left by the sandpaper action of boulders carried by the glacier don't begin to show up until about 900 million years ago. When they do appear, they mark the beginning of the longest sustained cold period in the history of the earth, one that lasted for over 300 million years.

That's the most bizarre fact about those glaciations: simply that they lasted so long. During the 300 million years or so of cold, several continents were marked with the scrawl of glaciers, and the cold appears to have come in phases: layers of rock that show evidence of glacial origin alternate with those that do not. Finally though evidence of glaciers from this period has been found from Australia to Greenland, the glacier-covered areas 900 million years ago appear to have been clustered fairly close to the equator.

This glaciation is just recent enough to have left sufficient detail to allow us to begin to draw a picture of climate change. Just as the earlier ice age implied, to keep the world cool for 300 million years required some significant alteration at the very heart of the climate system. Something had to have reduced the amount of energy from the sun reaching or remaining on the surface of the earth. At the same time, the stops and starts in the glacial record show that there must have been some climatic loophole: different parts of the world warmed up at different times, and the whole world, apparently, was not thoroughly chilled at all times.

The presence of variation within a larger pattern of major change leads Frakes to conclude that these glaciations probably had multiple causes, each of which may have prevailed at different times. Cosmically, the amount of sunlight reaching the earth could have dropped if the solar system as a whole were passing through a region of the galaxy that was thick with interstellar dust. Alternately, the earth may have been tilted over in its orbit at a much steeper angle than it is today, which would have altered the strength of the seasonal cycle. Earthbound, the position of the continents could affect when and where the glaciers occurred, as has been surmised about the earlier ice ages. The oceans are reservoirs of heat as well as of water, and drifting continents change the shapes of the oceans surrounding them; if the evolving geography of the oceans altered the circulation of water from the depths to the surface, the amount of heat energy stored or released to the atmosphere would change. Release too much heat and the changes in the oceans themselves could trigger ice ages.

The rocks again rule out some possible candidates. Increased volcanic activity, for example, is still an unlikely explanation, given the lack of volcanic deposits associated with the glacial remains. Common to all the conceivable candidates, however, is a message about the nature of climate change. The ice ages recur because there are, after all, a limited number of ways—mainly predictable ways—the earth can respond to a change in conditions. But most of these possible causes in themselves cannot be predicted. None of them need recur in precisely the same way; some climate changes, either those discovered in the past, or those we might yet experience, will always come as surprises.

Predictable unpredictability. The next ice ages add yet another twist to the ways in which climate can change. About 450 million years ago, in the geological period called the Ordovician, glacial deposits formed in North Africa; and again, 280 million years ago, during the Permian period, glaciers left their mark across much of the Southern Hemisphere. These two are the first ice ages for which a fairly detailed temperature record can be reliably constructed, using what climate scientists call evidence by proxy. They both occurred at a time when organisms were abundant and varied enough to leave behind fossils that could contain clues about the environment in which they had lived. The warmer a place is, generally speaking, the more types of plants and animals it will support. The greater the area that plays host to a wide variety of different species, the higher, on average, the temperature of the planet as a whole must have been. (There are plenty of other factors that affect species diversity, of course—humidity, the availability of nutrients, and so on—but the abundance of different forms of life does provide a rough measure of temperature.) Prior to the Ordovician ice age, fossils left by a number of different organisms were spread about the globe. On the basis of these fossils, the relatively widespread presence of reef structures, and other, more subtle geological indications, we may reasonably conclude that before both sets of glaciers began to spread, the world was a pleasant place, probably warmer than it is today.

How, then, did a relatively stable warm period shift to a relatively cool one? Again, continental drift seems to provide the most likely mechanism. Both ice ages occurred on landmasses that were then located near the South Pole. The more recent of the two probably began because the supercontinent dubbed Gondwanaland had been drifting rapidly, bringing more and more of its area close to the pole quite quickly. This meant that small, isolated glaciers might have formed over increasingly large land areas, eventually coalescing into the kind of ice sheets that exist today on Antarctica.

Ice sheets covering huge masses of land near the pole would, in turn, have helped refrigerate the planet as a whole, as the reflectivity of the ice again created a loop of negative feedback, bouncing sunlight back out to space, encouraging more cooling, more ice, and more reflection. The glaciers would have retreated, in this scheme, when Gondwanaland shifted its position, carrying some land further away from the pole. (The oceans also play a role here: just as one pattern of ocean circulation can lead to global cooling, so another can move heat efficiently from the equator to the poles, leading to a global warming.)

This is still all theory; attempts to simulate these conditions have shown that the theory is at least within the realm of possibility. But though the fossil evidence can suggest more or less what happened by indicating, for example, what now-separated landmasses were once joined and roughly when environmental conditions changed drastically, the presence or absence of a given set of organisms in the fossil record cannot explain how any of these changes occurred; causes are still obscure.

The Ordovician and the Permian glaciations did end, though, and what's striking is what comes next: good weather, 150 million years of it. Warm conditions were the rule from the end of the Permian through all the years during which the dinosaurs evolved and disappeared, and into the era in which mammals gained ascendance. Throughout this time the world apparently was free of large, permanent ice sheets, which means that its average temperature had to be some 10° Celsius warmer than at the present day.

Just as had been true of the ice ages, this prolonged warming did not leave any obvious clues as to its origins in the layers of rocks laid down during the onset of the warm spell. When working from first principles to try to account for such a warm period, we begin with the fact that climate was relatively uniform across the entire stretch of the Mesozoic era, the middle age of life on earth. To explain that kind of uniformity, climate theorists had to come up with some natural process that would have an impact easily maintained over the whole world. There is only one subdivision of the planet in which events readily, rapidly, and more or less evenly spread over the entire world: the atmosphere. Rocks build up and erode, the sea possesses its currents; everything on earth moves; but nowhere does novelty travel so easily as when it is borne on the wind.

So the task was to explain an increase in the world's temperature by identifying some characteristic of the Mesozoic atmosphere that is different from the air we breathe today. Framed that way, the question is almost too easy: A change in the amount of carbon dioxide in the atmosphere has to be the first suspect. A carbon dioxide–rich atmosphere absorbs and traps more of the heat radiated away from the surface of the earth, thus warming the entire earth, than an atmosphere poorer in carbon dioxide, which allows more heat from the earth's surface to escape to space, thus cooling the surface of the planet. But to argue that the Mesozoic atmosphere was richer in carbon dioxide than ours is just begs the question: If high carbon dioxide levels triggered and maintained a heat wave that lasted 150 million years, where did the gas come from—and why, in the end, did it stop coming?

The answer, or better, an answer, hinges on a willingness to predict backward. It is pretty well understood what controls the flow of carbon dioxide in and out of the atmosphere today; any explanation for the climate of the past that hinges on carbon dioxide assumes, of necessity, that the carbon cycle then behaved essentially as it does now. For some of the central details of the picture, certainly, that is a safe statement. In the past, as now, the earth's crust and its oceans have to be the major actors regulating the flow of carbon in and out of the atmosphere. Relatively speaking, the 750 billion metric tonnes of carbon in the present-day atmosphere is a trifling amount when one considers that the oceans hold more than 35,000 billion tonnes, and the crust of the earth contains 65 million billion tonnes more. In the long run, the amount of carbon dioxide in the atmosphere depends on how quickly the earth releases carbon to the air in a geological process called the "rock cycle."

Simply stated, the earth leaks carbon dioxide out through its cracks. The earth stores carbon in rock; its natural heat melts some carbonate rocks continuously, and as the rocks heat up, they degas, that is, they give off some of their carbon as carbon dioxide. Volcanic eruptions release the gas to the atmosphere as does the upwelling of magma along the midocean ridges, and in what one scientist calls the "Perrier effect," CO_2 bubbles up into the atmosphere through carbonated springs.

It was not until the early 1970s that any human being actually managed to reach the Mid-Atlantic Ridge or the East Pacific Rise, two of the earth's longest fissures; most of the two great seams remains untraversed. It is a landscape—a seascape—that is incredibly foreign. Black basalts form a crazy, jumbled ocean floor, and at the vents where fantastically hot jets of water shoot out into the depths, bizarre colonies of living things grow. The creatures that inhabit these areas tend to grow large: a shucked clam from the vents can weigh a pound or more. Crucially, these sites recirculate several compounds that had been trapped within the rocky interior of the earth, such as sulfur, which is the key nutrient that supports the vent communities, and especially carbon, released back into the carbon cycle.

The existence of such regions of undersea volcanism had been predicted since the late 1940s; the idea had emerged from a program of research designed to test the theory that the continents wandered—which eventually it confirmed. The discovery that continents drift had profound implications for understanding how climates can change on the moving landmasses, and it also helped to create another idea that linked the geochemistry of the earth (how chemical elements and compounds move from one place to another on the planet) with the broad regulation of change within global climates. The misshapen sculptures of volcanic rock at the ocean floor, the bizarre menagerie that makes its home within the islands of warmth created at the mouths of the deep ocean geysers—all these strange and foreign sights are pictures of the climate system at work. As they contribute to the deep-ocean carbon budget, they foster a multimillion-year cycle.

From the ocean to the atmosphere. Once the CO_2 bubbles up, it enters the other half of the rock cycle. As rock on land grinds slowly into soil, the weathering process produces silicon and carbon compounds. These compounds bind with CO_2 plucked out of the atmosphere to produce new molecules that dissolve in river water. The rivers flow to the sea; the carbon they bear reacts with elements in seawater to form calcium carbonate and magnesium carbonate; the new compounds float slowly to the bottom of the oceans, and there, over time, the sediment forms into rock. Finally, to complete the cycle, the slow, continuous movement of the tectonic plates that make up the earth's crust folds the rock on the seafloor down, deep enough into the earth to begin heating up the carbon they contain. As they warm, the carbon-bearing rocks bubble off their carbon dioxide all over again.

The whole process is extraordinarily slow: A single carbon atom could take perhaps 200 million years to travel through the earth to the atmosphere and back to the earth again. But although it is a ponderous vehicle, the rock cycle has at its disposal enormous amounts of carbon, which would enable it, in principle, to mediate great, gradual changes in the world's climate.

At last, we have enough real data to winnow out competing theories about what went on in the climate. To get Mesozoic global average temperatures some 10° Celsius warmer than our own as a result of a greenhouse effect alone, the atmosphere then had to contain between five and ten times as much carbon dioxide as it does today. The rock cycle certainly was revved up more during the Mesozoic era: painstaking reconstruction of the movement of the oceans has shown that the seafloor then was spreading apart more quickly than it does at present. When the seafloor spreads, new rock from the interior of the earth oozes out along the midocean ridge line, and the faster this new rock spills out, the more carbon dioxide the earth releases in the degassing process.

Evidence of ocean movement doesn't provide unassailable proof of a greenhouse effect, of course. Clearly, if the atmosphere today were suddenly to receive a huge injection of carbon from the interior of the earth, the world would warm; calculations of the likely impact of a higher carbon atmosphere in the Mesozoic era, all other things being equal, show that it was probably capable of producing temperatures around those required to produce an ice-free planet. The key phrase, though, is "all other things being equal." The best that can finally be said is that the idea of a prehistoric greenhouse effect does not contradict any of the hard facts left from the time.

In the absence of absolute proof, though, the idea that the Mesozoic climate was produced by a greenhouse effect lasting 150 million years gains support because it is parsimonious. It not only accounts for the evidence gathered from the field (the observed increased rates of seafloor spreading and the observed record of warm temperatures), but it offers a possible explanation for the larger problem of what regulates climate stability and change. The rock cycle provides one theory of how change on earth on the multi-million-year time scale can be both created and kept within bounds. From ice age to ice age and from balmy interlude to our own day, temperature has constantly varied, and yet the world has never fallen prey to a runaway greenhouse effect like that which has turned Venus into an oven, or like that which has settled Mars into an equilibrium of uninterrupted chill. With its steady production of CO_2, the rock cycle keeps the world from ever being completely covered in ice, because without exposed rock to weather and absorb CO_2, the gas would simply build up in the atmosphere until the world warmed enough to melt the glaciers. On the other hand, the continual weathering of exposed rocks prevents CO_2 from building in the atmosphere without limit, keeping a bridle on any greenhouse effect. At both climate extremes, the rock cycle is one of the key processes that enables the earth to support life. It allows the earth to become cold, but not too cold; warm, but short of boiling.

Back full circle: The rock cycle creates a world in which change is inevitable: the interaction of land, sea, air, and life is an endless accumulation of tugs, pushes, and pulls. The timing of any particular change, the direction that it might set for the world's climate, and the ultimate consequences of the new situation—all of these are much harder to predict. If for some reason the earth cools, the glaciers will probably spread, but for how long and how widely?

Every case is different. But anarchy has its limits: life and air, sea and land have between them evolved into a system that varies around a mean. Barring disaster, the system does not break down entirely. This, then, is the legacy of the effort to write the climate history of the earth. On the million-year time scale it is clear that global average temperatures can vary by as much as 15° Celsius or more; that the changes appear to occur relatively gradually; and that, while particular causes remain hidden, for virtually all that record the earth as a climate system has had a set of feedback mechanisms that keep the nature and extent of possible changes in check.

There is, though, one type of change for which the image of a hermetic system, a carefully tuned climate machine, makes no allowance. That is the idea of apocalypse. Sixty-five million years ago, a cataclysm occurred; at least one of the ideas about what happened seems most clearly to hinge on a single climate event. The Mesozoic era was the age of the dinosaurs; by the end of the era, they had evolved into the dominant animals on earth. And then they vanished swiftly, along with about a quarter of all families of organisms alive at that time. Certainly, if you take the dinosaurs' point of view, disaster had struck. According to the rock record, the temperatures before and after the great extinction remained much the same for several million years, but the biological disaster must have had a cause—and a particular type of climate change, a catastrophe rather than a gradual shift, could have been the culprit.

Actually, plenty of explanations have been put forth to account for what killed the dinosaurs. The Great Flood is one; another is that the dinosaurs were junkies, eventually succumbing to overdoses of psychoactive plants. Disney's animators seized on the idea of extinction as a result of catastrophic climate change: in *Fantasia,* the dinosaurs died of heat prostration, collapsing in a desert landscape growing ever hotter, ever drier.

In fact, the *Fantasia* scenario isn't all that implausible. There isn't much, if any, evidence to support it, but the cartoon version has this virtue: a worldwide weather change would explain the demise not only of the dinosaurs but of all other species that died out at the same time. The most popular scientific account of the end of the dinosaur era envisages just such a catastrophe of climate, one, however, that temporarily chilled the earth, rather than heating it. In 1979 physicist Luis Alvarez and his son, geologist Walter Alvarez, both of the University of California at Berkeley, proposed a radical theory that could account for such a swift, transitory shift in the world's climate.

Louis and Walter Alvarez began by wondering about an anomaly within the rock record that dates back to the boundary between the Cretaceous and Tertiary ages, 65 million years ago. Within these rocks, which are contemporaneous with the fossil record of extinction, is a strange band of material: a layer of isotopes of two rare elements, iridium and niobium. Although these isotopes are found in meteorites in the same proportions, they do not ordinarily show up except in minuscule quantities in the earth's crust. The anomalous band has been found widely distributed throughout the world, and it appears to have been laid down everywhere at about the same time. The only way such a layer could have formed, the Berkeley scientists reasoned, was if a dust cloud from a great collision between an extraterrestrial object—a giant asteroid or a comet—and the earth had settled more or less evenly over the entire planet.

The debris from such a collision, however, would first have risen into the atmosphere, creating a planetwide pall that could have, if the particles of dust were small enough, remained aloft for years. The Alvarez theory did not specify precisely how the dinosaurs (and the other species) could have died, but either the cold or the darkness produced in the shadow of the cloud might have been the culprit.

The idea itself was undeniably compelling, but what's more interesting is the relish with which it was taken up by both the scientific community and the public. There are other explanations for how the iridium-niobium layer formed. Certain types of volcanoes draw material up from deep within the earth, and the concentrations of the two elements in that volcanic material match relatively closely the levels found on meteorites, leading some scientists to argue that we need not look outside the earth for the iridium-niobium source. A more serious objection is that the dates of the mass extinction are not precise. Apparently some species died out before others, which reduces the odds that a single catastrophe was responsible. But the volcano advocates have not received anything like the publicity that the collision proponents have, and with refined versions of the theory to deal with objections like the problem of the timing of the extinctions, the Alvarez proposal even achieved the ultimate imprimatur: one extension of the original idea even made the cover of *Time* magazine in 1985.

On one level, the reason that the collision theory is so attractive is that it is simply exciting science. Ingenious, simple in concept, and at least plausible enough to argue about (after all, the great crater in Arizona, not to mention the pitted surface of the moon, provides mute witness that extraterrestrial rocks can strike home with stunning force), it explains as many, and in some cases more, of the observed facts as any of the rival accounts of the mass extinction. As had been the case with the rock cycle–greenhouse effect theory to explain the warm era that preceded the great extinction, simulations have shown that at the very least a massive collision between the earth and an extraterrestrial body could have had significant climate effects. While that doesn't prove that the dinosaurs succumbed to some kind of an "asteroid winter," it does show that such a winter could have occurred.

Even in the absence of proof, the immediate appeal of the

Alvarez theory demonstrated something of the nature of the change in the field of climate science. The asteroid impact theory cut across disciplines to resolve—or at least appear to resolve—outstanding problems in several of them. Such an interdisciplinary impact is one of the hallmarks of a new synthesis in science, and it is perhaps because of the simplicity and economy with which it simultaneously addressed issues in geology, evolution, and climate history that the collision theory, radical as it appeared on first reading, gained a significant measure of scientific credibility.

Extending beyond the realm of the professional scientists' debates, the theory commanded belief because it fit with what we are prepared to believe—we meaning everyone in our society, including, certainly, the scientists who conceived of it. Like everyone else in this last part of the twentieth century, I carry within my consciousness the images of mushroom clouds and devastating explosions; I remember as a child watching television shows about how to build fallout shelters, and I recall how the sound of a plane at night always made me wonder if this was the one carrying the bomb. We have all come of age with the certainty that when the world ends, death will come from the skies. The idea that the dinosaurs perished in the holocaust that followed the biggest impact of them all *feels* right because it fits so neatly into the nightmares that project our own demise.

Within climate science the impact theory is crucial because it completes an answer to the question with which we began: How can the world change, on any scale, and what determines the nature and the range of change possible on the planet? The essential message is one of stability, of gradual shifts. Changes in temperature occur slowly and are constrained within relatively tight bounds. Within that broad stability, however, exists the possibility of catastrophes. The dinosaurs are dead and our world now is cooler than theirs was. But even the sudden disasters, at least all those we have conceived of to date, have been kept in bounds: there has always been liquid water on earth, at least as far back as the oldest rocks can take us. The world has never boiled away its oceans, nor has it ever frozen them solid. Apocalypse there may have been; enormous destruction did occur; millions of living beings died; the dinosaurs are gone. But through all that devastation, three-quarters of the families of organisms on earth survived, and the broad conditions of climate remained much the same for millions of years after the dinosaurs' demise. If the climate system juddered under the impact of a giant comet, it swiftly righted itself and proceeded on its previous course: warm, but slowly cooling.

I've stressed order here more than chaos. It all depends on the lens you choose. At the greatest distance, the world moves with stately, fairly regular progress. Temperature dips and rises, but only within a narrow band. At closer range change becomes more anarchic, less immediately understandable. Why should the world have cooled some 60 million years ago, and not 65, or 70? Why by 10° Celsius, to pick a number, and not by 12° or 8°? And, back to the present, on this Boston winter's day chaos apparently reigns unchallenged: three inches of snow was predicted last night, but the ground is clear this morning. The weather pattern shifted slightly, and a storm that was right there, just out of sight, unaccountably vanished, surprising all of us here. But the absent storm is only a mild surprise: the underlying constant is that we are not now living in the peak of an ice age; there are no glaciers rumbling down Beacon Street despite the winter chill, and it is this fact that grounds the unpredictability of the weather within the broad stability of climate.

What remains is the problem of causes. We can propose mechanisms of change and believe them, more or less. We know that the continents move and that such motion must have an impact on climate; we know that carbon dioxide leaks out of rock; we know that debris from space sometimes strikes the earth: at any given time all of these factors have had some impact on the climate. More than that, for the distant past, for most of the earth's history, we cannot say.

And yet in the absence of means to create a detailed picture of the distant past, it remains possible to create one out of present experience. The Alvarezes came up with their idea because they were puzzled by some strange deposits in layers of rock; at the same time, a lot of other people were thinking thoughts that had little to do with the dinosaurs and yet prepared the ground for the collision theory. Thomas Pynchon's *Gravity's Rainbow* begins with the line "A screaming comes across the sky." He was writing about German V-2 bombs threatening London, but the sentence echoes within the mind's ear. The movie *Dr. Strangelove* showed up fairly often in Berkeley throughout the sixties and seventies, with its story of a doomsday machine. The ongoing (still, today) dispute about the MX missile kept the possibility of nuclear war on the front pages.

This is not to claim that the Alvarezes read Pynchon and came up with their theory. The point is that much of science cannot determine what is actually true, but rather must figure out what was most likely to be true—particularly in those disciplines that try to describe a sequence of events in the past, events the scientist can never witness. But what is most likely is a judgment call. We decide what is more or less probable based on our own experience, our day-to-day lives. Holocaust is an uncomfortably familiar concept.

An idea in science, as with ideas in any discipline, has to make sense on two levels if it is to gain acceptance. It must be "good science"—it must explain whatever set of data is at issue. But as science makes its way out of the lab and into a generally held picture of how the world works, it isn't enough just to be right: like Caesar's wife, it must seem to be so.

Perhaps the single central concept of climate science is the idea of feedback or, more generally, of connections between disparate places and processes, connections that go both ways. The temperature history of the world creates a vision of ponderously creeping continents, of the interior of the earth churning out gases quickly or slowly, of ice whose spread speeds its spread and whose melting encourages melting. The search for a single line of cause and effect is probably misplaced; in every case the dynamic appears to be one of action and reaction, of interaction reverberating all the way through the system. That the world is one world, bound tightly, is an idea that resonates. The evidence of climate science (the recognition, for example, that a meteorite's impact could cool the entire world, irrespective of boundaries) reinforces the popular view. Our experiences of the everyday (like that of the cloud of radioactive particles that escaped from the Chernobyl reactor, which brought home to an enormous public the connections that exist on earth, binding one place to the next) illuminate and are illuminated by the science that can dissect those ties and reconnect them, from the Ukraine to here; from distant time to the present.

CHAPTER 3

The Coming of Ice

Ice. I keep returning to the question of the ice ages, held by the image (fabulous, though I'm sure it is) of a wall of ice advancing through Canada and then south, herding human beings ever closer together, running us toward the equator. In one of his late writings, the naturalist Loren Eiseley quoted an Eskimo who said, "We fear the cold, and the things we do not understand."

Ice. The remnants of the last ice age persist. Greenland and Antarctica both retain continent-sized glaciers, giving us a sense of what the broader world was like when the ice flowed widely. The ice has come and gone across the continents several times over the last five million years, or so the evidence of the rocks indicates. But in Antarctica there is ice that has remained frozen for half a million years, providing a continuous record of ice hundreds of thousands of years long. The ice is like a vault filled with the fragments of ancient texts, there to be reassembled by anyone with the wit to ferret them out. This ice—actually, particular ice cores drilled in Greenland and Antarctica—contains the earliest direct record of climate and climate change during the ice ages. In it can be found the clues that suggest how the world has cooled and heated during times recent enough and swiftly enough to touch the edges of human history, and to reach into at least some kind of human memory.

The last ice age ended approximately 12,000 years ago. That, though, was simply the most recent retreat in the tally of the forward and backward marches of the great sheets of ice that stretch back several million years at least. In the context of the great glaciations, like that during the Ordovician period, we are still in an ice age. Shift again on the time scale, focus in on periods of tens of thousands of years instead of millions, and what now becomes obvious is variability, change, chance. Boston is ice-free now, yet was not just a brief time ago, geologically speaking. But what is visible is not simply change but patterns within the apparently random flux of change, of heat or cold. Each great glaciation of geological antiquity possessed its own unique history: particular processes triggered them, like the movement of continents, and events particular to each shaped their courses and brought them to their end, all within the broad constraints of the evolving climate system. Our own ice age is different, or at least near enough to our experience so that we can perceive the differences. Crucial cycles link the apparently ungovernable ebb and flow of the ice. Some physical mechanism drives the variation we can observe in our own climate regimen, and the search for this apparently fundamental property of our planet's behavior became one of the central quests of the researchers trying to build an interdisciplinary science of climate.

The history of that research in fact marks out clearly the point at which the tools of the discipline gained the power to resolve the issue, creating in the process a new view of the world, of the planet as a whole. There is now an elegant explanation that accounts, more or less, for the pattern of the ice ages, one that has emerged in pieces over the last three-quarters of a century, coming to a state of relative completion only within the last three or four years. The story of how that explanation came to be is at once a story of the triumph of this emerging science of climate and, in its final twist, a cautionary tale as well.

Over the last several million years the back and forth sweep of the ice has followed a pattern that has been mapped within the rock record. The largest shifts in ice sheets have followed a 100,000-year cycle, but shorter fluctuations also have taken place. Reasoning from fairly obvious first principles, the regular variability in the amount of ice on earth indicates regular changes in global average temperatures, from which it follows that the balance of energy absorbed and emitted at the surface of the earth has to be changing in step. From these premises it is a small leap to suggest that an economical solution to the problem of what triggers changes in the amount of ice covering the earth would lie in a mechanism that could alter the amount of solar energy reaching the earth in a regular fashion: the less sunlight, the more ice, and vice versa. Following this path of reasoning leads to a fact of orbital physics well known in the nineteenth century: The earth leads a merry dance around the sun.

As the earth travels its orbit, it tilts at a slight angle and spins around its axis like a top. That spin has a wobble in it, called "precession," and each wobble takes place over 22,000 years. Another jitter in the earth's spin repeats itself every 19,000 years. At the same time, the angle of the axis itself is changing, from a minimum of 21.5° off the vertical position to a maximum of 24.5°. (It currently leans over at a 23° angle.) The tilt swings back and forth to a 41,000-year beat. As the world wobbles and dips, the amount of energy each hemisphere receives from the sun changes, one gaining and the other losing some warmth. On an annual basis, this orbital geometry creates the pattern of the seasons, summer coming when our hemisphere is tilted in the orbital plane toward the sun, winter when a half circuit leaves us tilting away. Over millennia, the theory goes, the alterations could create the climatic analogue to seasons: ice ages and interglacials.

The idea that these relatively small changes in solar energy reaching the earth could cause the beginning and end of ice ages did emerge during the nineteenth century, but none of the early schemes was able to match the timetables of the orbital variations closely enough with the observed cycles of the ice ages. The problem was finally solved when a Yugoslavian mathematician named Milutin Milankovitch calculated the effect on temperature that each fluctuation of the orbit could cause and then laid the precession effects atop the changes caused by variations in tilt of the earth's axis. Milankovitch argued that summertime temperatures were the most crucial measure because a cooler summer would mean less of the winter snows would melt. From thirty years of calculations, he concluded that the drop in the amount of sunlight reaching the earth at the extreme end of both the precession and the tilt cycles could cool a hemisphere enough to set the ice on the march again.

Milankovitch began winning converts to his theory as early as the second decade of this century. However, it wasn't until 1976 that (temporarily) final proof emerged. This proof turned on a separate discovery, that the sediment that builds up on ocean floors contains an index of the world's temperature on time scales short enough—again, a few thousand years—to reflect the cooling and warming that attend the movements of the ice. Part of the sediment consists simply of dust, minerals, and clay, but part of it forms out of the continuous rain of the shells of tiny animals called "forams," or just "bugs" to the researchers that work with them. The chemical composition of foram shells varies with the conditions within the ocean. The differences in question here are very subtle—a cool ocean contains a very slightly higher concentration of the heavy isotope of oxygen O^{18} than does a warm one. It wasn't until the seventies that the tools became available to measure the tiny variations in the heavy oxygen content of foram shells. The climatologists Nicholas Shackleton, James Hays, and John Imbrie examined several cores of sediment taken from the deep ocean, including one that stretched back over 450,000 years. They found that the temperature dips and hence the ice ages occurred regularly, a few thousand years after each peak and valley in the cycle of the earth's tilt and wobble. While this did not fully account for the 100,000-year cycle, all the shorter swings in the ice matched up with the orbital variations that moved to beats of several tens of thousands of years. All other things being equal, the next ice age should begin in 2,000 years or so, with a slow cooling trend that would eventually set the glaciers off again.

A triumph—a virtuoso, classical demonstration of the scientific method. Milankovitch provided the hypothesis and a theoretical explanation for the events of nature; Shackleton, Hays, and Imbrie performed the crucial experiment. The hypothesis was confirmed, neatly, and an observed pattern of climate change had its explanation. But there's something more here, more than just the technical mastery the scientists displayed. The kind of explanation Milankovitch offered, together with the elegance of the experimental confirmation, provided almost a propaganda victory for the emerging science of climate: not only were the questions interesting, but the new science could generate answers that not only seemed right but also possessed a quality that workers in other fields could recognize as beauty.

Science gets a bad rap because the math is hard, the technology daunting, and the work—digging about in the ice or running a deep ocean drill off some workaday ship—is so prosaic. But ideas in science are as much a creative product as they are in a painting or a book, and they can be judged by the same criteria. Do the parts of the reasoning fit together nicely? Is the explanation complete? Does the argument advance with grace, without too many stumbles to account for stray details? Does the idea possess that wonderful clarity that makes it seem like it ought to be true?

For climate science, emerging in this instance out of a marriage of usually quite disparate fields—astronomy, physics, oceanography, geology, and meteorology are all represented in the ice age problem—the attribute of beauty was very close to essential. The idea behind climate science is one of unification. Using a single line of reasoning, drawing on a host of different kinds of information, researchers had the task of seeking out the simplest set of connections that bind the earth's history to its current state. In a sense, climate science had to come up with something more than simply a solution to a problem. An ugly idea, convoluted, complex, dependent on one random accident of nature or another (not interstellar dust or drifting continents again!), provides no incentive for someone trained within a particular discipline to broaden his or her focus. The attraction of an interdisciplinary field is the promise it offers of cutting through problems, to take Alexander's approach to complex knots, but to succeed it must deliver on that promise. The clean, simple lines of Milankovitch's idea (for all the incredible laboriousness of the actual mathematics) were a triumph of aesthetics, one that enables this new science to say, "See—we bring order out of nature." Beauty alone won't do it: Milankovitch's final vindication rested on the count of oxygen atoms found in the shells of innumerable microscopic animals. But at least part of the reason that his theory lasted long enough to be tested against the impartial testimony of the forams is that Milankovitch had created an argument, an explanation, a good and satisfying story about how the world works—a solution that can be acknowledged as beautiful.

That solution was complete and generally accepted by 1980, at the latest. But then, almost overnight, it was challenged by the physical testimony of ice that remains from the last ice age, and ice beneath it that was laid down by the last interglacial, and so on, back through several turns of the Milankovitch cycles. That ice contained within it artifacts preserved intact from the climate of the ice ages. The evidence of previously unsuspected changes in the chemistry of the atmosphere during the height of the glacial periods forced an immediate reexamination of the Milankovitch theory. Over the longer term the evidence from the ice cores has forced the new science of climate to the recognition that the climate system is even more delicately interwoven than had been suspected.

Unfortunately, the ice in question hardly has the aura one associates with the high drama of a scientific revolution. It's not the Rosetta Stone by any means, nor does its importance leap out at the casual observer in the way the moons of Jupiter must have stunned any Renaissance man brave enough to bend to the eyepiece of Galileo's telescope. Simply getting the ice cores out of the glaciers requires more simple, brutally hard labor in daunting conditions than it does any great burst of intellectual inspiration. The longest cores, the only ones so far that have been drilled clear down to bedrock, came first from the Camp Century site slightly to the north and west of the center of Greenland, then later from Byrd Station on Antarctica, and finally in the late seventies and early eighties again from Greenland, at the Dye 3 radar station, south and east of Camp Century. One measure of the battle involved is that the Dye 3 drill site is abandoned now, unfit for use, though the drill there only touched bottom in 1981. On Greenland the ice age has not ended to this day: the snows of winter don't melt and the glacier grows a little every year. The radar station itself is mounted on pylons, and becomes a kind of raft that floats on the ice; the drilling station, in contrast, was anchored directly in the ice and has gone the way of a sand castle facing a rising tide.

The Greenland coring expeditions were joint projects, mounted by American, Danish, and Swiss scientists. They were inspired by the International Geophysical Year (IGY) in 1957–58. The IGY was a rather quixotic idea, whose aim was, in essence, to find out as much as possible about what the actual nature of the earth was in a twelve-month span, with the ultimate hope of fitting all the pieces together over time. The first short cores were drilled then, but it was not until Chester Langway, now the senior American ice miner, went to Camp Century with a new drill in 1961 that it became possible to reach back deep into the ice-age record. The work was confined to the summer months, and Langway depended on Dartmouth students to supply the grunt labor. That involved wrestling with the drill itself and handling the sections of the core, columns of ice five inches in diameter and several feet long. It took five years to drill to bedrock, more than a mile down, but by 1966 Langway had a column of ice that formed a continuous time line from over 100,000 years ago to the season he started drilling.

At his lab at the Buffalo campus of the State University of New York, Langway presides over the American stockpile of polar cores—more than 30,000 feet of ice. Most of the cores remain on the seventh floor of a high-rise commercial refrigerator downtown, but Langway keeps a few hundred feet of ice handy in a small vault within his lab. The cores themselves have an alien look to them— you would never encounter anything that has precisely the appearance of a polar ice core anywhere else. They are heavy, cumbersome things, more than 6 feet long. They are dully translucent, and the light seems to sink into the center of the core; they have none of the sheen of the thin skins of ice that float on ponds in wintertime. Langway makes absolutely sure that the cores never melt: they are kept in a cold room chilled to −35°C—so cold that when you move from the storage area into the workroom, cooled merely to a balmy 10 degrees below zero, your glasses fog and you feel quite warm.

Some of the cores have been beveled down, different researchers having cut samples to perform one experiment or another, leaving the core looking like a log in the middle of being hewn into planks. Some of the effort goes simply to dating the core, trying to count each year's layer of accumulation. Other experiments include looking at particulates that get trapped in the ice, tracing, for example, the history of above-ground nuclear testing by measuring the presence of radioactive dust within the core. Langway husbands the ice very carefully; the three deep cores make a finite resource, and every experiment uses some of them up, irretrievably. During the sixties, a profligate time, ice (like a number of other resources) seemed endlessly available; when a Swiss physicist named Hans Oeschger joined Langway on the Greenland glacier in the middle of the decade, looking for clues about the ice age atmosphere, he used a ton of ice for each experiment. By the seventies, he performed the same tests gram by gram.

It was Oeschger who in 1979 managed to pry the telling clue about the ice-age climate out of the ice. In a crucial series of experiments performed in the late 1970s, Oeschger cut a series of small samples from several lengths of the Camp Century ice. He then crushed each chunk and collected the gasses that had been trapped in air bubbles throughout the ice. When he and his colleagues tested ice formed at the height of the last ice age, about 17,000 years ago, they found that the carbon dioxide content of the ice-age atmosphere measured about 180 to 200 parts per million. In samples taken from ice formed after the world warmed again, between 10,000 and 12,000 years ago, carbon dioxide levels leapt to between 260 and 300 parts per million.

Epiphany. Oeschger and a group of French scientists repeated the measurements on cores drilled near Byrd Station on Antarctica, and then from a number of other sites on the great continental glaciers, north and south. Each time they achieved the same results: during the ice age, the earth's atmosphere contained less carbon dioxide than it did a few thousand years later when the ice sheets shrank and the world warmed. Thousands of years now, instead of billions or millions. The ice ages have a special power to awe because they happened so quickly, geologically speaking. Oeschger's find provided the crowbar with which to pry apart the system and uncover the best picture yet of how climates change fast.

Once again carbon dioxide provides the key. Swift changes in carbon dioxide levels imply that something was affecting the rate at which life was consuming oxygen and respiring carbon dioxide. The orbital variations, which certainly occur, have no direct connection to the rate of change in the production or consumption of chemical compounds in the biosphere or within the atmosphere. Apologies to Milankovitch, but a change in the chemical composition of the atmosphere does demand some mechanism that couples orbital mechanics with the behavior of biological systems—a much more complicated problem than that of the physics of orbital motion. Oeschger's find immediately posed two questions. First, did the drop in carbon dioxide levels simply reflect the presence of an ice age—that is, was the fact of cooling because of Milankovitch planetary motions sufficient to drive levels of a greenhouse gas down? Alternately, was carbon dioxide a transmitter and amplifier of the orbital effect, and, if so, what was the connection between the carbon content of the atmosphere and the Milankovitch orbital cycles, themselves undeniably present in the record of the ice ages?

Even before Oeschger published his work, some nagging details about the timing of ice ages had begun to gnaw at a few scientists, prompting them to look for a process that could work alongside the Milankovitch cycles. Significantly, no variation in the earth's orbit can explain the 100,000-year-long cycle of the ice ages; this longest and strongest of the temperature cycles seemed to follow a beat of its own. And just as strikingly, there is an awkward syncopation to the rhythm: the earth appears to remain cold for a long time, and then warms rapidly, abruptly, with the cycle beginning again with a long, slow cooling trend. Why the sudden blast of heat (sudden meaning perhaps over a few thousand years in a cycle that runs for 100,000 years)? The initial explanation was that the 100,000-year cycle and the pattern of recovery reflected higher-order consequences of the observed cycles; that is, when the effects of the 19,000-year, 23,000-year, and 41,000-year patterns combined in a planet that has given properties of heat retention or radiation, they would jointly produce this strong 100,000-year signal. That argument depended on complicated calculations, and it seemed both incomplete and unsatisfying—difficult to defend.

The ice core bubbles marked out what to look for to escape the difficulty. Shackleton and a group of scientists reexamined the deep-sea sediments to find out if the oceans preserved a record of atmospheric carbon dioxide that matched the direct evidence in the ice bubbles. They found it possible to assay the carbon content of the surface waters of the ocean by comparing the ratio of different isotopes of carbon in the shells of surface-dwelling forams with those of bottom dwellers. That ratio provides an indirect measure of the amount of carbon dioxide present in the atmosphere. That led to two discoveries. First, the deep-sea sediments confirmed the ice bubble results: carbon dioxide levels in the atmosphere did rise and fall, according to the forams, when the sediment temperature record showed warmings and coolings; and at the 100,000-year interval, carbon dioxide appeared to leap, abruptly, and then sink toward a more normal level. Even more neatly, the shifts in carbon dioxide always occurred after the orbital juke associated with a particular warming or cooling in the spread of ice, but before the actual temperature rose or fell. If true, these results implied that carbon dioxide did in fact transmit, and possibly amplify, climate changes set in motion by a relatively small change in the amount of solar energy reaching the earth.

If true. I've been using the word *fact* quite freely, and perhaps a little misleadingly. It sounds as if the facts are always there, waiting to be found, like placer gold in a western stream. But theories determine their own facts; each new version of how the world works forces a rearranging of the observed details into a configuration that wasn't there before. The deep-sea sediments are solid, heavy, often messy agglomerations of silt and clay and bits of animals, as factual as you can get. But what they *mean* depends on how you look at them. In this case, study of the ice cores first had to produce the idea that the carbon dioxide record revealed an underlying process. But the ice cores revealed nothing about the nature of that process or mechanism. So how do you judge what the significance of all this might be? In this experiment, you see how consistent the relationship between carbon dioxide and temperature is, and then you see if the observed relationship fits a persuasive theory. If it doesn't—well, perhaps what you're looking at is an artifact, a mistake, one of nature's jokes played upon her inquisitors.

The task here was to build a theory that incorporated a plausible picture of what biological communities would do in response to changes in the amount of sunlight reaching them, with a plausible model of ocean and atmospheric chemistry. The first synthesis came from the ocean chemist Wallace Broecker, a researcher on the staff of Columbia University's Lamont-Doherty Geological Observatory. In this case the relevant living communities had to be ocean life, plants, and algae. While the oceans contain far less carbon than does the earth's rocky crust, the amount of carbon in seawater is more than fifty times that contained in the atmosphere, more than enough to act as a sink or source for the observed shifts in the core samples.

In principle, the idea of the oceans as the regulating valves controlling the global carbon cycle opened up a very simple scheme for Broecker. Carbon dioxide enters the sea from the atmosphere by dissolving into the surface waters; when the water is saturated with carbon dioxide, the atmosphere stops transferring carbon to the ocean, and any extra carbon dioxide remains in the air. Plants growing in the ocean consume carbon, creating more room for extra carbon dioxide in the atmosphere to dissolve into the sea. More plants, Broecker reasoned, would consume more carbon, so more carbon dioxide would be able to move from the atmosphere to the ocean, which in turn means that less of the gas would remain in the air, which finally would mean that the world would cool, in a mirror image of the greenhouse effect. Fewer plants, and the process reverses: the ocean would absorb less carbon dioxide, more would stay in the atmosphere, the world would warm, and the ice sheets would retreat.

That's the basic idea—and it is the limit of scientific advance at the time I write this. A half dozen different schemes explain just what regulates whether or not the plants in the ocean flourish, but none of them has been proved. My favorite, again on aesthetic grounds, is one developed by the Harvard climate scientist Michael McElroy, together with his student Fanny Knox Ennever.

McElroy noticed that the densest plant growth in the ocean occurs at high altitudes, near the poles. So he argued that the plants there would especially thrive whenever changes in the earth's orbit increased the amount of sunlight they received. The lush growth would fix carbon from the seawater, drawing more carbon dioxide out of the atmosphere, cooling the earth, following the regular rhythm of the shorter Milankovitch cycles. Finally, every 100,000 years, the ocean would suffer a kind of population crash: the life in the ocean would outgrow the supply of oxygen available to it. With the collapse, the ocean would absorb negligible quantities of carbon dioxide, levels of the gas would rise quickly in the atmosphere, and the earth would warm, setting the stage for another 100,000-year loop.

Broecker himself and at least two other groups have come up with roughly similar ideas. They argue that the circulation of ocean water at high latitudes changes during ice ages. In the cold periods, the circulation would carry more nutrients—nitrogen and phosphorus, for the most part—to the surface, fertilizing the plants there; during the shifts to a warmer climate, ocean circulation would revert, so fewer nutrients would reach the plants at the surface. All of these theories have a common drawback: Evidence from the deep-sea sediments does not show the changes in the nutrient and oxygen levels in the sea when each of the theories predicts they ought to be there. As Broecker says, "Sometimes when new evidence comes along it doesn't immediately solve any problems—it just shows that the problems are more complicated than you thought."

What comes next? New information. Again, this is an example of the intellectual feedback loops that drive science. Theory, the competing ideas that McElroy, Broecker, and the others have developed, is forcing a hunt for new clues about how the climate of the ice ages actually behaved. The nuggets of fact, as they tumble out of the masses of mud and ice, in their turn rewrite theory, until the gap between what the theorist expects to find and what the world actually contains narrows enough to persuade the scientist that his solution is basically true. For the ice-age question, the next crucial data will probably come from an ice core that Soviet researchers are drilling at Vostok Station in Antarctica. It is already the longest core ever drilled, and when the Soviets reach bedrock, they should have a continuous record of half a million years of ice—long enough to compare directly with the full extent of the deep-sea sediments.

It's easy to look back into history and say, "This was a turning point; this marked the time from which change was inevitable, when the revolution triumphed." Events that at the time seemed trivial, ordinary, unremarkable grow in history till their significance to an observer is obvious, with the perfect wisdom of hindsight. In the middle of the game calls are harder to make, but I'll gamble: Oeschger's discovery of low carbon dioxide levels in the atmosphere during the last ice age has transformed the world.

That discovery was so crucial because it was one of the last nails in the coffin of the clockwork universe. The world Oeschger's findings replaced was one that behaved rather like a child's marble, spinning about its orbit according to a short and simple list of rules. McElroy recalls the days of box-and-arrow diagrams, where the carbon cycle, for example, is reduced to a catalogue of carbon reservoirs (the boxes) with carbon moving along the fixed tracks laid down by the arrows. Climate, ice ages, any fluctuations here clearly are results—the consequences of some external push or pull, the flick of a shooter's thumb, some agency from outside the cycle. It's a clean image, with cause and effect neatly separated.

The new world—well, it is a much more uncertain place. What the carbon dioxide results do show is that the simple box-and-arrow models of the earth are far too static. The crucial cycles that determine what kind of climate the planet may enjoy are far more malleable than people had realized before the direct, unequivocal facts about the ice age atmosphere forced them to think about such things. Change the system anywhere along the cycle, rev up volcanoes or starve life in the oceans, and the entire system will judder, adjust, and produce a new climate regime. And that change will in turn reverberate through atmosphere, ocean, land, and life, keeping the cycles of change rolling merrily along.

I'm deliberately overstating the case. The idea of a static world began to die in the nineteenth century, with discoveries like that of Louis Agassiz and his original conception of ice ages. Darwin helped as well, of course; the concept that the forms of life are themselves changing constantly has been one of the most powerful corrosives eating away at the old world.

What the carbon dioxide discoveries do, in a sense, is to bring Agassiz and Darwin together—and here the matter of time scale is crucial. Take a long enough view, then all that matters is physics. Over tens of billions of years, the sun flares up and then dies; its planets spin on their orbits; their cores cool; the radioactive elements within them decay; and eventually they become inert—little balls in space, marbles. On the billion-year scale, life and the physical environment interact in a process of slow evolution, establishing ultimately an apparently stable environment in which new life continues to emerge. What change occurs in the environment seems simply to spur evolutionary processes, a simple, unidirectional path of cause and effect. Continents move, oceans change shape, and species alter.

By the same token, look too closely, and you see accidents and chance events come into focus, but there is no sense of a larger pattern linking one process with another. On a single day a storm dumps its rain on the land below; the seasons change as the earth travels through each section of its orbit; some years are drought years, some flood. All the events on this scale affect life on earth, but over the short term there is not a whole lot that plants can do about it, except to adapt to the constant reality of winter and summer, the existence of storms, the risk of drought, or whatever else may change in their environment.

But life and the weather do intersect; the orbital variations Milankovitch found, the carbon dioxide levels Oeschger measured in the ice cores, and the tracks of repeated ice ages in the record of the rock all are intertwined, forming a world that transforms itself constantly. The importance of Oeschger's work and of the research that has followed to explain his findings is that it has clearly shown that the behavior of living systems affects as well as is affected by the physical conditions in which the life in question finds itself. What plants do is both an effect and a cause of swift climate change; that makes the climate system, the metabolism of the earth, as it were, much more complex than it ever seemed before.

Climate science is new, born in part of this problem and this resolution, these disparate lines of research. And it is the new knowledge that the physics of planetary motion and biology of life in the sea can interact to produce regular climate change that transforms the world within the mind. I said earlier that what was destroyed was an image of a world with an unchanging history. That is not quite accurate, or not quite complete. What changes with the climate as well is an image of a world that will always remain the same. Just as the weather we have today has not remained constant in the past, there is no reason at all to expect that conditions of climate that we have grown used to need remain the same in the future.

Ice. I am fascinated by ice, held by the images of woolly mammoths and ancient man, fleeing south or making, like the Inuit today, some terrible, harsh accommodation to lands grown white and desperately cold. When Louis Agassiz (with some assistance, certainly) "discovered" the ice age, that astounding revelation propelled him into the first rank of science's heroes. But then he lingered on, moving from Switzerland to Cambridge, Massachusetts, where he ruled over Harvard's biology department and ultimately became the leading American opponent to Darwin's theory of evolution. If science really were the bloodless pursuit of truth, achieved by piling one discovery atop the next, then it would be immune to tragedy. But ideas have life spans of their own, and they can, when the time comes, leave the scene grudgingly, carrying with them into irrelevance all those people who cannot bear to give them up. Agassiz's ultimate fall into a peculiarly precise brand of creationism overlaid with a nasty streak of racism has the air of tragedy to me, as much as any story of some God-maddened Greek king.

By now, Agassiz is long dead and mostly forgotten. He survives as a name on one or two buildings and in a couple of pages in histories of science. He does leave behind, though, an object lesson about the stakes for which science and scientists play. The stories scientists tell are not simply bedtime tales. They place us in the world, and they can force us to alter the way we think and what we do. To Agassiz, Darwin's new world was no fit place in which to live; he failed to accommodate himself to it, and he was doomed.

And a world in which ice comes and goes according to the multilayered rhythms of planetary motion and the carbon cycle has dangers commensurate to those that Darwin's visions possessed, at least for some. The old world—the world we lived in perhaps a decade ago—was simple enough to have some rules to live by. The forces that dominated it were the large ones: the rock cycle, the movements of the continents, and the wobble and roll of the earth in orbit. Human beings couldn't do much about these sources of climate change; they just happened, independent of any human (or other living) intervention.

Easy street: we go our way, and the earth keeps on its own path. The new world does not permit that kind of detachment. It's more complicated by far; the rules are less obvious; the risks are greater. We know now that the conditions in which we thrive can alter as a result of subtle changes that seem at first glance to have nothing to do with the average temperature over the latitudes in which we live. With slightly increased sunlight over the arctic seas, plants could thrive and thus draw down carbon dioxide out of the atmosphere; then, in principle, we could freeze. Take the argument one step further: things we do that seem terribly far removed from questions of climate could in direct, physical ways transform our world into a place that becomes increasingly hostile. Although the mutability of this new world clearly makes us vulnerable, it remains fairly easy for us now to continue to blunder through the mass of connections that makes up climate, laying the ground for changes we cannot predict. For example, carbon dioxide levels in the atmosphere have been rising for more than a century as humans burn fossil fuels; given the role of that gas in regulating the ice ages, we can expect something to happen to the weather we experience, but we cannot be certain precisely what the change will involve, what cost it will impose on us.

Ultimately, the new climate science will reduce that uncertainty. After all, that is its goal, to identify with increasing precision how the climate system responds to different stimuli. But any new knowlege, any new world, carries with it a price. The increase in our understanding of climate has brought with it a burden of responsibility to monitor ourselves and understand what the risks are of any action we might take; we cannot any longer rest blithe and innocent of the consequences of what we do.

CHAPTER 4

Hard Times

THOUSANDS OF YEARS, centuries, decades. Life and climate intersect, but the nature of the meeting changes on each time scale. People experience climate as weather. It snowed six inches in Boston last night, but under overcast skies the temperature crept up just above 40°F by noon today, and the gutters are ankle deep in slush at this particular moment. On a little larger scale, every winter in Boston seems unspeakably harsh to this California native, but I've lived in the Northeast long enough to be able to tell the difference between bad and worse; compared with the blizzard year of 1978, this season has been beautifully benign.

Weather is obvious. You know when it snows, you can feel the change that comes with the spring thaw, and when it rains, your feet get wet. At some level climate is as easy to comprehend: everyone grasps the fact that some winters are harder than others; some springs short, and some long; some summers blisteringly hot, and a few (usually the ones longest ago, at the edge of memory) are absolutely perfect.

But I cannot tell, just from my own experience and recollection, whether today is an especially unusual day or a normal one well within the accustomed range for a given season. If we have a string of very cold or surprisingly warm days, I don't have any way to judge if this represents just a cold snap or a heat wave, or if it is a symptom of climate change. I recall a sequence of three years in the seventies when it seemed as if the East Coast got nothing but snow and more snow, but the number of years I have to measure them against is too small for me to say whether the series of bad winters was just coincidence or a harbinger of more to come. By the very end of a life it becomes just possible to judge, to say, "Yes, those were hard times," or to recognize the sequence of fine years, but in the middle of actually living through good days and bad, there is no way to say for sure whether the weather over the long term —climate—is getting worse or better.

It is a time scale that creates a set of problems for science as well as for memory, presenting difficulties that do not arise when the issue is either the daily weather or a climate record stretching over thousands of years. In fifty years, or a hundred, average temperatures can vary by as much as a degree or two Celsius—a large enough difference to have a profound effect on the weather that the people living through the shifts must endure. But there is no clear-cut mechanism that drives temperature change on this almost-human scale that can match the elegance of the Milankovitch effects and the carbon cycle regulation of ice ages. Sun spots could play a role, perhaps, though no significant one has yet been found for them; shifts in the circulation of air through the atmosphere may be crucial, or perhaps atmospheric changes are the result of temperature changes, and so on. Within this brief a span it's almost impossible to decide which events are parts of cycles, connected in a chain of cause and effect, and which are simply linked by a chance juxtaposition.

The issue becomes almost preposterously complex, though, when we add to the problem of the causes of short-term climate change the question of what impact that change has on the course of human events. Where previously events occurred in evolutionary time, spans long enough for life and climate to interact, now we have entered historical time, in which the dynamic is not one of interaction, but of action and response, of a cold winter that living things must endure, well or ill as the case may be. The science of the climate of decades and centuries becomes, at least in part, a science centrally concerned with definitions: What is the climate regime of the moment? What drives it over brief periods? How changeable is it? Even if all these questions were easily resolved, the outstanding issue of how human societies recognize and respond to climate would remain, does remain, as one of the central foci in the study of the natural history of our species. The climate record—

written in ice cores, in the remnants of fossil plants, in the advance and retreat of glaciers in the Alps—and the history humans have made and written down seem to run on parallel tracks, never touching. But climate must make a difference in the way people lead their lives; contrast a Californian with a Bostonian for an obvious (if fairly trivial) example. So the question becomes, what difference has it made? Did Napoleon fail in his invasion of Russia because of the vicious winter, or did the winter merely hasten a failure made inevitable by Napoleon's faulty tactics and the endless harassment of Russian armies, which enjoyed practically unlimited room in which to retreat and maneuver? I believe in the second answer, but no absolute proof decides the issue either way.

The problem is that it is impossible to design an experiment that would isolate precisely what happens at the intersection between nature and human affairs. The number of variables that affect any given outcome, like Napoleon's triumph (occupying Moscow) and demise (retreating from Moscow), is too huge to handle in any simplified model, like those used to reconstruct prehistoric climates. And then, because the data of climate are historical (that is, the events in question have already occurred and cannot be precisely duplicated), even the ordinary logic of laboratory experiments won't work. It is impossible to go back in time, change just one parameter, and see what happens to the rest of nature and to the (human) history being made. What is possible, in fact what is the only point of entry into the whole question of how climate and human history intersect, is to examine the records of both and seek the coincidences that may be more than merely coincidental.

Human history, as opposed to prehistory, begins with such an apparently meaningful juxtaposition of events: the ending of the last ice age and the shaping of human settlements. From that time (about 12,000 years ago) to the present day, the average temperature of the world has ridden a roller coaster. The climate optimum, the warmest, most congenial time the world has known recently, occurred between 10,000 and 11,000 years ago. A crucial time: Just about then human societies began to make significant use of agriculture, and hence began increasingly to divide labor up within groups, to organize, and to amalgamate. Weather obviously matters to farmers, and good weather makes it easier for beginners at the game to make some headway, but there is no way to establish a cause, only a correlation between the fact of increasingly mild conditions and this fundamental turning point in human history.

The climate optimum did not occur everywhere at the same time. In North America, the glaciers persisted, even advancing slightly about 8,500 years ago. Indians in what is now the southwest United States remained hunter-gatherers until about 6,000 years ago, when, in a period during which Arizona and New Mexico were wetter than they are today, local peoples began to raise food crops as well. The world cooled a bit after that, beginning just over 5,000 years ago with a chill that lasted a few hundred years, judging from the telltale jumbled rocks that mark the advance of small glaciers left behind in the mountains.

After another glacial expansion just under 3,000 years ago, the world then began to warm, through the time, as it happens, that the legendary Yellow Emperor brought rice culture to the Chinese. Mild weather persisted. The weather was warm in Europe up through the Middle Ages, warmer than it has been ever since. The height of the extraordinarily easy conditions, called the Medieval Optimum, came between A.D. 900 and 1000. That was the time of the Viking conquests, with dragon ships ranging as far south as the Mediterranean and as far west as North America. It was a time when feudal fiefdoms began to coalesce into nations, more or less: the Danish kingdom forming in the North; France beginning to unite in a process that lasted several hundred years; and the English monarchy establishing itself in 1066 with the Norman Duke William's victory over the Saxon Harold.

William, now King William, moved rapidly to establish Norman power over a recalcitrant Saxon majority. He ordered a census of his new realm, the results of which were recorded in the Domesday Book. Among the tallies of fields and structures and serfs, one list leaps off the page—a holding that doesn't belong in England, at least to the modern mind. The Domesday Book identifies thirty-eight vineyards, with grapes growing as far north as York producing wine considered by the census takers as good as the French vintages the Normans had left behind them. For Yorkshire to support grapevines, the average temperature there would have to have been between 1° and 2°C warmer than a Yorkshireman can expect today.

The grapes of Yorkshire died out long before the present,

however. The Northern Hemisphere cooled, beginning in about A.D. 1300. From 1550 up to as late as the nineteenth century the cooling in some places was severe enough to again set mountain glaciers on the march, fast enough to swallow up tax-paying farms in alpine valleys. This period, known as the "Little Ice Age," was marked by shorter, wetter summers, at least in some parts of Europe, and winters that were on average longer than those the continent had endured in medieval times. Growing seasons were reduced by as much as a month, according to agricultural records of the day. Pollen records show the boreal forests retreated to the south throughout the hemisphere. The English vineyards failed and sea ice became an increasing problem for northern harbors.

And people starved. Hard times—hard for people, an identifiable point of intersection between climate and human society—recurred throughout the Little Ice Age, most obviously on the margin between ice and the first hardscrabble villages, but throughout Europe as well. In 1565 the grain harvest in upland Switzerland was disastrous; the crops failed again the next year, and then again in both 1571 and 1573. The winters had been unusually long in those years, with snow cover lingering into the late spring—forcing farmers to dip into hay stocks to feed their cattle. When the hay ran out, they had two options: feed their herds on pine branches or slaughter them. The long, cold springs also nurtured a parasite that attacked the grain fields. One bad year could lead to hardship, but not penury; it was the relentless march of one harsh season after the next that transformed bad times into crisis.

Perhaps crisis is the wrong word—it implies that famine or the threat of it was somehow unusual as well as dire. I think it is a conceit of our century that famine seems more of an isolated problem than it used to; it is something that happens to faraway countries, not to us, people living in the advanced countries of Europe, North America, the Pacific Rim. But from 1600 to 1783 a contemporary author counted forty general famines that swept through all of France, one of the richest and strongest powers in Europe. Climate historian Christian Pfister counts fifteen catastrophic harvests in Switzerland between 1565 and 1824; ten of those occurred within climate conditions more extreme than any that the Swiss have had to deal with in the twentieth century. Hunger, starvation, and disease were routine visitors to the towns and farms of Europe up to the very edge of the modern era. The conclusion seems obvious: the climate downturn in the middle centuries of our millennium was sharp enough to impose great hardship on the human beings and societies forced to endure the chill.

One last coincidence, however: From 1500 to 1850, throughout the Little Ice Age, the nations of Europe expanded in population, power, technological competence, military strength, economic endeavors, in world rule—in virtually every measure of the vigor of a civilization. At the end of the period one nation, Britain, ruled a quarter of the globe (and bought its wine from France.) The climate change occurred, no question; it caused privation that periodically included outright famine and death by starvation. But amid the tangle of events that add up to the history of that time and place, it is impossible to determine what difference the Little Ice Age made to the history of Europe; impossible to predict Europe's destiny if the weather had remained mild; impossible for the scientist to unravel the historian's mysteries.

The upshot is that climate science, for all its remarkable ability to reconstruct in detail the history of climate over thousands of years, cannot do for human spans of time what one expects a science to do: explain the underlying machinery that drives events. Simply knowing that the Little Ice Age occurred or that during the 1950s much of the world enjoyed an especially benign climate regime does not even begin to answer the central questions. What drove the change? The events move too quickly, climatologically speaking, to offer any obvious mechanisms for the shift. What did the change do, in practical terms, to the people on the ground? What did the changes mean in history? At any given point, it is impossible to break into the affairs of human life to isolate one event that occurred because of the weather (unless it be a game at Fenway, called on account of rain). Climate science finds itself, even with its impressive armory of tools and technique, forced back to an older tradition of thought: understanding nature gives way to the simple task of describing it.

This older form, natural history, is foreign to ideas of science formed with the example of laboratory studies in mind. For climate science, the world can be seen as a laboratory in the round, with natural events as experimental results. The work of the scientist, then, is interpretation, to reason back from data, from the event in nature to the initial experimental design that could have produced that data. Nature itself is almost an irritant, a screen placed in the way, obscuring the underlying machinery of cause and effect. For children and a few adults this schism between nature and science does not exist. In E. B. White's *The Trumpet of the Swan,* Sam, the boy naturalist, takes the measure of the world around him in "a cheap notebook always by his bed." At the very beginning of the narrative, Sam describes the scene that foreshadows all the adventures to come:

I saw a pair of trumpeter swans today on a small pond east of camp. The female has a nest with eggs in it. I saw three, but I'm going to put four in the picture. I think she was laying another one. This is the greatest discovery I ever made in my entire life.

Among the great adult naturalists, perhaps John Muir best combined exuberance in the face of the unrelenting detail of nature with careful precision in his descriptions of what he saw. In one passage, Muir depicts the particular type of winter storm needed to create the long, whirling contrails of snow off mountain peaks he calls "snow banners": "Many of the starry snowflowers out of which these banners are made fall before they are ripe, while most of those that do attain perfect development as six-rayed crystals glint and chafe against one another in their fall through the frosty air and are broken into fragments." Classic Muir: meticulously observant—banners are made from snow dust produced by the destruction of proper crystals; and poetic, even romantic— "starry snowflowers," a vivid image, a phrase that locks into memory.

But science remains natural history for just a brief time. Natural history is seen as the pursuit of amateurs like Audubon and Muir; for the professional scientist, careful description is the point at which work begins, and poetry is usually as foreign as the sun worship of the pagans. Science is supposed to provide meaning, to determine what is and isn't significant, what actually happens. One graduate student I know is spending six to eight months, forty hours a week, processing dirt dug up from a lake bottom. When he's done, he should be able to construct a time sequence of rainfall for the last several thousand years, but he won't even be able to begin to do what one ordinarily thinks of as science with this material until he has given over better than half a year of his life to the task of looking at mud through a microscope. For my graduate student acquaintance, his lake sludge is just that, sludge, until he can identify what is in the mud and compare one layer with the next.

The discipline that provides the sharpest sense that it is science that gives meaning to nature is physics, with its search for laws and unifying organization. But the harsh truth is that the central concerns of physics are too remote for most purposes. "A theory of everything," now being sought, can explain nothing interesting, if what interests us includes a level of detail messy enough to include the vagaries of human life. Climate as a system is vastly more complicated than physics; even as a formal problem, purely within science, describing the rules that underlie the establishment of the layers and regions of the atmosphere, movement within that structured mass of air, and interaction between it, landmasses, and the ocean, it presents an enormously more difficult challenge than any one posed by particle physics, for example, or cosmology. Add the element of human affairs and the best that we can do is to seek within the welter of climate records and human history specific incidents that can offer a vantage point from which to view the connections that may exist between events, natural and human. When the question is, what does climate mean to man, the answer is that it means different things to each man and woman in each place, at each time. But if the question is what *can* climate mean to man, then it is possible to find examples from within human experience at the extremes of climate conditions that provide a point of entry.

There is a story of the Little Ice Age, a tale from the edges of Europe, that gives not the answer, but an answer. It begins, as a good saga ought, with murder. In Iceland, sometime around A.D. 980, lived a man named Eric, styled Eric the Red. Eric was born of a family used to mayhem; he and his father removed themselves to Iceland from Norway to avoid the consequences of their having committed manslaughter there. Once in Iceland, Eric killed two men in a feud and was banished to an outlying district for his pains; then more men died in a dispute over property. When the feud threatened to expand into a pitched battle, the Thornessthing (one of the Norse courts on Iceland) outlawed Eric and his supporters.

Eric's enemies then began to try to track him down among the known islands of the Norse world. After he had completed the rigging out of his ship for a long voyage, Eric told his friends that he intended to sail west, away from the known world, to try to find the land an earlier sailor had seen after being blown off course. Then (according to tradition the year was 982) Eric and his household sailed to a landfall at a place called Blacksark, an ice mountain. From Blacksark Eric turned south, searching for habitable lands, and eventually found a network of fjords reaching far into the interior; along the fjords lay land good enough to support the Norse. Eric wintered in the new territory, then sailed back to Iceland. After losing a battle to his old rival, Eric and his enemies resolved their feud, and the next summer Eric led an expedition to colonize his discovery, which he named Greenland because, records the saga, "Men would be more readily persuaded thither if the land had a good name."

According to the saga, Eric was something of a swindler, for by the time history was being written, Greenland was rarely green, and Eric's promises must have seemed like those of the smooth-tongued men in new suits who sell waterfront property within the Florida swamps. The fact was that the interior of Greenland always belonged to the ice, but Eric had landed at the height of the Medieval Optimum; the climate, though harsh, was just mild enough to permit the Norse to support themselves. The Norse settlers clung to a series of ecological islands along the fjords where the ocean currents kept the temperatures moderate enough to enable the colonists to pasture cattle in the summer and raise enough hay to keep their herds alive through the bitter winters.

Actually the settlers were able, for a while, to do more than just squeak by. They formed two separate settlements, one around the southern tip of the island, and the other about 250 miles north along the western coast, with a combined population of about 4,500 people. In addition to their cattle, sheep, and goats, the Norse supported themselves with an annual seal kill and hunted other arctic animals, caribou, polar bears, and walruses. They even maintained a sporadic trade relationship with the European mainland, exchanging walrus ivory and hides for iron, wood, and church ornaments. By some reckonings, the Greenlanders may even have seemed rich: in 1125 the community traded a live polar bear for a Norwegian bishop and built him a farm and a cathedral on some of the best land in the whole colony. There was enough of an economic surplus to support a church-building program that left Greenland with several of the largest stone buildings to be found anywhere among the North Atlantic islands.

Although the hunting was still good enough in 1127 to enable the Greenlanders to pay a crusade tax of nearly 1,400 pounds of ivory, the Medieval Optimum in fact ended on the island before it did in Europe. At some time beginning around 1270, the climate faced by the Norse turned increasingly harsh. Within a century or so the increasing cold apparently made an already harsh land virtually uninhabitable. Ocean storms began to reach up into the inner fjords. By 1350, the smaller settlement along the western coast had been abandoned; the amount of sea ice clogging the sea-lanes increased. The last bishop of Greenland died in 1378, and trade with Denmark and Norway dwindled in the fifteenth century. By 1500 every Norse settler on Greenland was dead or gone.

It is a tragic story, but a good parable. In this case the climate indisputably played a crucial role in the death of an entire, if small, society. With hindsight, it is clear that the Greenland Norse were virtually always at risk. Even good years must have been good only at the cost of the extraordinarily hard labor needed to bring in sufficient surplus so that people could make it through the winters. The great hunts were communal, and, according to climate historian Thomas McGovern, every seal kill strained the resources of at least the poorer households in the colony. As generation succeeded generation, always just on the edge of safety, the odds must have increasingly tilted in favor of disaster.

That's true enough, and it's possible that the colonists themselves might have seen their fate in those terms. But there is one fact left out: At the same time that the Norse vanished from Greenland, the native Inuit hunter-gatherers persisted and survive to this day. According to McGovern, the Inuit succeeded where the Norse failed simply because they bent to the wind, where the Norse did not. The Inuit followed the food supply; the Norse stayed with their farms, and good land was increasingly concentrated in the hands of fewer and fewer families and the Church. When the harp seal catches declined, the Inuit hunted ring seals and caribou. Inuit harpoons were superior to the nets the Norse used; Inuit boats made from skins were better for coastal transport than the wooden craft of the Europeans. The Norse even persisted in their taste for woolen clothes, tailored to a European cut; the Inuit wore garments sewn from animal hides.

The Norse colonists are a terribly remote people now, and their deaths a distant sorrow. In one sense they leave a strikingly obvious lesson behind them: they so thoroughly failed to adapt that it's almost enough to conclude that they shouldn't have been so inflexible. If anything, the fate of the Greenland Norse proves not that climate determines human affairs but that climate affects a people only as much as they let it. That's an exaggeration, of course, as much as the idea that the Little Ice Age alone caused the demise of the Greenland colonies. But strip away the drama and romance of the rise and fall of an embattled colony of settlers, and the broader meaning of the story begins to reveal itself. The simplest moral to draw is that climate does not just happen to people in the same way that weather, a storm, for instance, does. When it rains once, you find your feet have gotten wet; when it rains all the time you learn how to build leakproof roofs. What human beings experience as climate is the product of a relationship between the actual climate conditions, be they optimal or as chilling as the Little Ice Age, and what human beings do about them.

That concept is illustrated clearly in the history of the vanished Norse colonies; it is harder to trace the connection between human and natural events in places and times when the consequences of error are not so severe or apparent. What the Greenlanders demonstrate is that climate change always forces a choice on the societies experiencing the shift; and that any choice has a cost. The Greenlanders hung onto their culture. They remained herders and occasional hunters; they stayed fixed to farms and their stone churches; they wore clothes their cousins on the mainland would have recognized; and eventually they died. The other option was to abandon all or part of their trappings of European life, emulate the Inuit, and, perhaps, live. A bitter choice. In more complicated histories, in richer, more secure places, the welter of details obscures the starkness of the choice, but the parable of the Greenlanders reminds us that there is always some price to pay to maintain a particular way of life in the midst of conditions of climate that never remain absolutely the same. We may not be able to recognize the coin, or even calculate the specific price we pay; the dead Norse force us to remember, though, that pay we must.

And the role of science in all this? Science diminishes the specter of chance, caprice, pure accident in human affairs. With the historian's advantage of hindsight, the clear physical evidence of cause and effect provided by a phenomenon like the Little Ice Age cuts like a scalpel through the obscuring welter of detail that any human society leaves behind it. The last strand of meaning that emerges for me from the story of the Greenlanders is that even though climate reconstruction does not and cannot provide a complete explanation for the fate of the Norse, it provides at least a place from which to begin to understand the events. More generally, even when science fails to answer a specific question, the central tenet of the scientific faith, that material events have material causes, always holds out the prospect that an explanation exists, even if the murk of daily life obscures it.

Compare Eric's saga with a story, a much older one, in which climate change plays a major role. In it the pharaoh of Egypt dreamed strange dreams, that seven sickly cows devoured seven fat head of cattle, then that seven healthy ears of corn sprouted off a single stalk, only to be destroyed by seven ears "thin and blasted by the east wind."

It was Joseph, Jacob's son, who interpreted the pharaoh's dreams. He foretold seven rich years and seven lean to follow, and he advised the pharaoh to choose a man to look over Egypt, appoint overseers, and hoard one-fifth of the produce in the good years "that the land not perish through the famine" to come. And the story concludes with the triumph that always thrilled me: the pharaoh appoints Joseph to run Egypt.

I have just now read those passages of Genesis again, for the first time in a while, and there is more to the story than I had remembered, layers of meaning that run beneath the obvious messages in the text. The conflict in the story seems to run between Joseph and pharaoh, then between Joseph and his brothers, who fail to recognize their kin in the potentate who stands before them.

But a line of tension between the land and those living on it also runs through the story. The threat to Hebrew and Egyptian alike at this juncture comes not from men, but from the famine, the blasts of the east wind that scour the corn and starve the cattle. The underlying current is a crisis brought on by catastrophic changes in the daily conditions of weather and climate and how the people living in the blasted lands accommodate themselves to conditions they cannot control: "And all the countries came to Egypt to Joseph to buy corn, because the famine was sore in all the earth."

In this story the word of God removes the terrifying sense of the arbitrary from life, and faith provides the ability to accept even the less palatable events that might flow from divine commands. We ask of science something similar: to produce a sense of order within inchoate experience; to explain; perhaps even to deliver on the promise of power, the ability to use the knowledge gained of order and the mechanisms of nature to create a world that pleases us. And science, in a broad sense, has seemed capable of doing all that. The world today is vastly different from the world of a couple of decades ago; it is changed out of measure from that of my parents' childhoods. Science as an article of faith—the faith that material events have material causes—is inextricably intertwined within the transformation that has occurred, now driving it (with the enormous impact on belief, say, of relativity and quantum mechanics) and now driven by it (as, for example, with the development of the atom bomb and the invention of the transistor, both applications of modern physics, carried forward by history).

And yet, with these monitory tales from Greenland or a vanishingly distant Egypt, climate science offers a warning, or at least a sense of limits. The faith that material events have material causes is a treacherous one: it does not affirm that those causes will be knowable in every case. Climate in nature is complex enough that at certain time scales—like those of a human life—the causes of most events are inherently obscure. We know that climate matters to us. We can take the reconstruction of Greenland's rise and fall as part of the record of medieval Europe and read into it a parable, written by scientific research, of the potential impact of climate on human affairs. But it remains a parable, not a prescription. There is no magical escape built into our growing understanding of nature. We know that the climate we possess now, in the latter part of the twentieth century, will alter, and will do so within our lifetimes. I know that this will have some impact on my life, whether or not I discern it; what lesson I draw from history, that of the earth and that of its people, is that I, and all of us, must be prepared to respond to events we cannot predict and cannot change.

CHAPTER 5

Hard Years

I~N~ DECEMBER 1982 the rains fell on Louisiana as if they never meant to stop. The Red River overflowed its banks; so did some of the streams to the east. In the small towns near them, people rowed their boats through the streets. Louisiana was flood-covered once again, and the state waited and watched, maintaining a vigil lest the Mississippi itself breach the walls built to keep the great river in check. Rain fell and fell some more; it seemed as if the water might rise from Christmastime till Easter.

In 1982 the sun shone as if it had outlawed rain. Indonesia suffered a drought so strong that the rain forest in Borneo dried out and then burnt, and forest fires swept through 10,000 square miles of East Kalimantan. When the monsoon failed over southern India, reports circulated that in some places water sold for a dollar a gallon. In Australia they called it the "Great Dry"; instead of rain, Victoria and Southern Australia suffered storms of dust.

It was a time when the weather seemed malicious no matter where you looked. The Galápagos Islands, usually dry at that time of year, had plenty of rain. Tornadoes struck southern California; Tahiti endured typhoons. Perhaps Peru had it worst: the rains there were so severe that whole hillsides—and farms—washed into the rivers and out to sea.

To the citizens of Louisiana the events of the winter of 1982–83 were clearly defined. A series of storms kept coming in from the west, and they brought a lot of rain; when Louisiana gets a lot of rain, the rivers rise past flood stage, and people get their feet wet.

In India the central element of human existence that season was the fact that the sun kept on shining. In Australia, the sky grew dark; there is a famous photograph, looking for all the world like a WPA image made during the Dust Bowl years in Kansas, of a wall of dirt bearing down on the outskirts of Melbourne. But people lived through all this weather, through the heat, the dessication, or the water falling in sheets.

And it was more than just people that suffered. Whole ecosystems changed. The nesting grounds on Christmas Island, temporary home to millions of birds, remained essentially empty for a season. Seals raised litters of pups on the Galápagos Islands, but then abandoned them when the food supply ran out; pelicans left their fledglings to starve on the Catalina Islands off southern California. Again, Peru suffered inordinately. The catch from the anchovetta fishery, which fell by half in the early seventies, dwindled to almost nothing in 1982–83; when the fish died, both gulls and fishermen went lacking. Floods stand out; thirst stands out; hunger leaves an indelible memory. There is no question here of slow climate change, of the undetectable fluctuation of a degree or two from the average temperature in a lifetime, causing a swing from easy times to hard. This was disaster, devastation, whether a person lived on a parched farm in India or one washed away in Peru; no need here for any argument about the subtle interaction of climate—or at least a season of terribly bad weather—and life.

Time, again, defines the issue, time and now distance. Over great stretches of time it is easier to trace broad relationships of climate the world over—between the amount of sunlight reaching the earth and temperature, between events in the ocean and the atmosphere. In the space of a year the tie that binds conditions in Madras to those in Louisiana is more obscure. Peru is impossibly distant from Indonesia, from California—at least it seems so when the weather of the day is a rainstorm that one man watches as his own field turns into mud and then vanishes. The existence of a catastrophe that reaches swiftly across so many miles begs the same question as do the longer, more ponderous climate events: What within the system can account for such change? The problem is expressed in terms that are perhaps the oldest and most basic in all of climate science: What will next year be like? Is there anything about this year's weather that can presage the weather twelve months hence? The first answer is that the events of 1982 are all consequences of one turn of the climate machine, a turn a little more violent than most, but one that is recognizable and for the first time understandable within the terms of the new science of climate.

That cannot be said for many climate disasters. Sometimes what happens simply happens, for no good or predictable reason. A year is simply a bad year or a good one. A unique, unpredictable catastrophe can pass for the cause: the explosion of the Indonesian volcano Tambora in 1815 threw between 24 and 48 cubic miles of rock into the atmosphere as dust, creating a veil that remained aloft for more than a year. The veil cut the amount of sunlight reaching the earth and was blamed (possibly mistakenly) for the frequent frosts that struck throughout the succeeding months; in New England the chill was such that 1816 became known as "the year without a summer." More immediately, I can state with absolute certainty that the weather a year from today will be different from what I see now, looking out my window, even though I can do little but guess what that weather will be.

But there are also climate events that are not random. Though it is scant consolation to someone wading through a flood that used to be a farm, some of the cruelest weather falls into patterns—skeins that can be unraveled, examined, perhaps reveal a sequence that can be anticipated. Some ties do bind across half a world. As distinct from each other as they seemed, the events of 1982 were bound by such ties and, taken together, form a single pattern. They are, in fact, the most dramatic recent example of a class of phenomena that occurs around the globe. The underlying processes act as switches, altering the basic climate pattern across a large region, and it is those mechanisms embedded in the climate machine that form the connections between drought in Australia and rain in Peru. Simply uncovering those links, just describing in detail what happened, was one of the great triumphs of climate science, one that took the better part of a century to achieve.

In Peru they have a name for what happened in 1982–83: *El Niño*, "the Christ child." El Niño is a familiar visitor in Peru, arriving around Christmastime, hence the name. In most years what the Peruvians call El Niño is just a current of warm water moving southward from Ecuador to the northern coast of Peru during December and January, bringing a characteristic pattern of sea breezes. Full-scale El Niños, the strange and dangerous events that include both warm ocean currents and several months of heavy rains, recur every three to eight years.

As far as Peru was concerned, until very recently El Niño was a familiar, unwelcome, but tolerable companion. It has been a feature of the climate of the area for as long as anyone has been able to check: there are records dating back to 1726 that mention El Niño's effects, and it has been suggested that the heavy rains carried by one El Niño made it possible for the conquistadores to cross the forbidding desert regions of Peru a century earlier. Traces of ancient floods in Peru's Moche river valley suggest that El Niños, some much stronger than any recorded ones, have occurred frequently for at least the last few thousand years. What caused them remained a mystery, but the phenomenon seemed purely a local problem.

The monsoon in India—steady winds that bring with them the regular annual rainy season wet enough to support rice cultivation—has a recorded history that stretches even further back in time. References in Western writings to the monsoon and the prevailing winds that accompany the rains are found in sailing manuals dating from 1584; departures from Goa back to Portugal or on to China were timed to take advantage of the season. Within India the agrarian economy was balanced on the climate pattern, so precariously so, in fact, that caprices of the weather translated directly into fortunes of men and women. Just five years after the catastrophic failure of the monsoon rains triggered the famine of 1899, Gilbert (later Sir Gilbert) Walker became the director-general of observations for the colonial government of India. He had read the work of some earlier British scientists who had measured a seesaw effect in barometric pressure readings: when pressure was low over India, Walker knew, it was high over South America, and vice versa. Walker took that piece of data and spent thirty years measuring and counting to add detail to the picture. He was able to show that the pressure shift correlated with anomalies in weather all over the globe; most important, he was able to prove that droughts in India were linked to the flip of the seesaw that sent pressures down over the southeast Pacific and up in the west, over Australia and Indonesia.

Walker was by all accounts a remarkably thorough man; what he saw in the record, the correlations he pieced together from all the fragments of information from weather station after station, did exist; he documented a genuine physical process. But Walker had to stop there. He had no theory that could explain why pressures in the southeast should matter to atmospheric pressures in the southwest, nor why the shifts he certainly found should affect the weather the way they did. He simply described the world as he found it.

There is a curious fact about Walker and his work. He was recognized as an unusually meticulous scientist, admired for his ability to pick out the threads that belonged in the pattern he created while discarding all the dross. And clearly the research he was doing was of double importance: it was a marvelous piece of science, essentially discovering a whole range of physical interactions between dramatically separated parts of the globe; and it was of obvious human relevance since people suffered when the world behaved in one way, enjoyed richer times when it did not. But for all Walker's honors and the significance of his research, his results essentially languished for forty years. Perhaps Walker was almost too good at what he did. Mark Cane, one of the scientists currently studying El Niño, thinks so and suggests that Walker's lack of theory, lack of an explanation for what he found, created a loophole that allowed other researchers to ignore Walker's information. Other scientists may have taken the very comprehensiveness of his stacks of data as a sign that the effects Walker found were simply accidents of statistics. The patterns Walker saw, by this line of reasoning, were patterns of numbers only, as if Walker had studied ten years of roulette results and published statistics on the number of times an even number came up immediately after the wheel had hit a zero. The data would be real, but its predictive value would be nil.

More to the point, El Niño was still too unwieldy a problem to tackle. The technology did not yet exist to process all the data properly, and even after Walker's work, it was still very difficult to see the details of what actually happens in the air and in the oceans when the zones of high and low pressure shift. It was possible to observe the beginning of Walker's sequence of events with the shift in pressure and then to observe the endpoint of the failure of the monsoon, but the intermediate steps remained obscure. The next step came out of a truly obscure field of study: the meteorology of the Marquesas Islands.

John Leighy of the University of California was the researcher, and he found that when India was wet, when the monsoon arrived on schedule, the Marquesas Islands in the mid-Pacific were dry. Then he noticed that under these normal conditions, when barometric pressure was high in the east and low in the west, the trade winds blowing east to west were strong. Finally he argued that when the winds were strong, the temperature of the surface seawater would be lower than when the winds were weak. This left Leighy with a chicken-and-egg problem: because the sea surface temperatures and the strength or weakness of the winds clearly varied in conjunction with each other, he couldn't distinguish which of the two was the primary factor that determined whether the Marquesas Islands would be wet or dry. Still, Leighy's work, published in 1933, was the first to build on Walker's findings by pointing to a link between atmospheric circulation, ocean temperatures, and the weather. It was a remarkable leap in awareness of the complexity of the climate system, but even more than Walker's, Leighy's work disappeared into limbo for two generations. The next paper that cited the Marquesas study came out in 1978.

The synthesis that was needed to put all these various pieces of the weather system together into a single coherent theory turned on a piece of luck. By chance a major El Niño occurred during the 1957–58 International Geophysical Year. For the first time the ocean warming off Peru was monitored deep into the Pacific. The data turned up that year provided the last clues that meteorologist Jacob Bjerknes needed to come up with a model that could explain how the whole ocean-atmosphere system works.

Under ordinary conditions, Bjerknes argued, when the great barometric low is located over Indonesia and Australia, the western Pacific is warm, and that warm water maintains the powerful center of rainfall in the west that India experiences as the annual monsoon. By contrast, the sea surface off the coast of Peru during this period is relatively cold—it is made up of water from the deep ocean that wells up to the surface near South America. The temperature difference between east and west affects the atmosphere as well: the sea surface temperature gradient drives air along the ocean surface from the cooler east out to the warmer west—the trade winds. The winds in their turn set up a feedback loop, pushing warm surface water along in front of them, piling up warm water in the west, moving it out of the way of the upwelling cold water in the east. Thus the winds maintain the conditions that set them blowing in the first place.

The spread of warm water to the east in 1957 showed Bjerknes how things could go wrong. If warm water extended eastward, the rain would also move east; the temperature difference between west and east would drop in turn. The Pacific low-pressure zone would now be found to the east of its ordinary position over Indonesia and Australia; the winds would weaken, allowing even more warm water to flow east and slowing the rise of cold water from the deep. Feedback again, working in the opposite direction. As long as warm water continued to spread over the eastern Pacific ocean, Bjerknes said, the El Niño pattern would maintain itself: it would be dry in the west and wet in the central Pacific and the Americas. As long as those conditions continued, the Pacific low-pressure zone would remain displaced to the east.

Success. Bjerknes's idea was a classically elegant example of scientific insight. With a single economical mechanism, Bjerknes bound together a range of disparate observed events, ones occurring in different places and in the different media of water, air, and land. There were still some problems with the theory. Bjerknes lacked a triggering mechanism, something that would explain the shift in the Pacific low or the initial spread of warm seawater. And that gap in turn created the other major difficulty: how the feedback mechanism established by the strength or weakness of the trade winds could ever permit the ocean-atmosphere system to switch from ordinary weather to El Niño or back again.

Still, Bjerknes's basic picture has held up, and his link between atmospheric and ocean events remains the best explanation of the physical process that produces El Niño. Since his time, other researchers have found similar systems at work in other regions: one associated with a low-pressure zone usually found near Iceland, for example, and another in the south Atlantic, whose fluctuations have been linked to the cycle of drought in central Africa (as have those of El Niño itself). These ocean-atmosphere/surface temperature-weather variations, these links, have been dignified by the name "teleconnections"; as such they are recognized as a novel kind of geographical feature, new lines with which to draw a map.

Here lies the surest measure of the accumulated discoveries that culminated in Bjerknes's studies: by this work the world has changed. We have seen it change, transformed by the recognition that Madras is tied to Lima, Jakarta to New Orleans. We have a new way of looking at the world, one that sees it not as a collection of separate, clearly identifiable pieces in a puzzle—here an ocean, there an island, next a continent, and above it all the air—but as an interwoven net of currents and winds, water evaporating from sea surfaces and condensing again as rain, a system whose boundaries blur and fade and always extend past the arbitrary limits of shoreline, nation, and hemisphere.

And yet Bjerknes's accomplishment, which represents the culmination, the apex, of a half a century of effort, is one of the last products of an older tradition in climate science. His success, in fact, laid bare the limits of this kind of science. Bjerknes took the evidence of meteorology, oceanography, and atmospheric physics to develop a theory to explain how the parts function together as climate. But that theory in its turn posed the next set of questions, and these Bjerknes, and the traditional disciplines, could not answer.

The change has been between science that describes a process and that which is concerned with explaining the underlying machinery that drives the phenomenon under scrutiny. Bjerknes laid bare what was the underlying process that produced El Niño events. The next task was to provide a dynamical picture of that process, one that could use observations of a given set of conditions in the Pacific to say whether they will interact to produce either El Niño or a normal year. The facts alone are not enough to explain what happens in El Niño events or in climate generally. Leighy, for example, knew that both sea surface temperature and trade winds were different at the Marquesas Islands during the anomalous periods, but even with Bjerknes's account the question of how the shift from one set of conditions to the next occurs remained unanswered.

The issue remained open for another generation. When the 1982–83 El Niño occurred, it provided a fairly dismal picture of how few new developments in the field there had been since Bjerknes. That El Niño was itself a double anomaly; it was both the alternate, less common prevailing weather system and a most singular and unusual version from a class of similar events in that climate pattern. It caught observers by surprise, but it provided enough new information to help generate the first plausible science that incorporates Bjerknes's picture into a theory that can explain the dynamics and open up the possibility of predicting the course of future El Niños.

Actually, on one level, the 1982–83 event followed a familiar course. The low pressure swung eastward, as did the warm surface waters; the trade winds failed and the weather patterns followed the usual El Niño scheme, if with unusual intensity. But all these phenomena were noted after the event had already begun; at the time, the 1982–83 El Niño took everyone by surprise.

Certainly, it was far stronger than any previously recorded El Niño: at its height, the seawater off Peru was 18°F warmer than usual, an almost unbelievable jump. The rains were more severe; the drought was worse; the impact on marine life and humans alike caught all the species concerned off guard. Most striking, however, the event occurred out of season. It is clear after the fact that El Niño began sometime in the late spring or summer of 1982, several months earlier than the conventional pattern would have predicted. Observers in the central Pacific noticed a rise in sea surface temperatures over the summer, but it wasn't until September and early October that the first telltale signs of El Niño registered off the Peruvian port town of Paita. By the time that the scientific community recognized that a full-scale, apparently quite large El Niño was under way, the damage that it could do was already beginning to happen; the rain had begun to fall, the flood waters were rising, and in the west, in Australia and India, the fine, clear weather was about to become drought. For all that had been learned about teleconnections and the progression of El Niño events, this particular disaster confounded climate science; in its timing, severity, and persistence it simply caught everyone by surprise.

It should have been possible to do better. Although Bjerknes's theory did not allow for forecasts, a certain kind of prediction could have been made. Once the first hints of El Niño displayed themselves, once warm water shows up surprisingly far to the east, for example, then in theory it would have been possible to have extrapolated from previous El Niños and to have given people likely to be affected a warning of the range of weather they could expect. Since 1982–83 Eugene Rasmusson at the National Oceanic and Atmospheric Administration has monitored, very much in the style of Gilbert Walker, any evidence he could find of sea surface temperature changes in the mid-Pacific in the hope that he will be able to recognize the characteristic shape of the most embryonic El Niño. Ideally, this could lead to predictions as much as a year and a half before the weather associated with El Niño actually made its presence felt.

That would be virtue enough, clearly; with eighteen months to prepare you can do at least something to protect yourself from drought or to shore up your defenses against flood. Fishermen can make plans against the probability that they will lose at least a season's worth of fish; governments could, at least in theory, actually plan ahead for something that had previously fallen into the category of an act of God, which could only be suffered and endured. Clearly, if we can get a jump on the weather by a season or two, our lives will be made easier.

Unfortunately, though a number of researchers including Rasmusson have explored this kind of descriptive prediction, no one has actually done it yet. It may, however, prove to be almost impossible to distinguish early on the genuine changes, the real march eastward of some warm water from the ordinary ebb and flow of ocean waters or random variations of the wind. If so, that leads to a second issue, one that at first sounds almost like an aesthetic objection to the whole notion of the kind of inelegant grubbing after facts required by Rasmusson's approach.

Basically, it would be gratifying if someone could actually predict El Niño before it happened at all. It would be more than pleasing actually—an understanding of the internal dynamics of the ocean-atmosphere system that produces El Niños is the key to determining one of the fundamental properties of the earth's climate: namely, how far in advance, and for what parameters, it is possible to predict the future of climate by examining the present. And El Niños recur with a kind of regularity and in a pattern that repeats itself over the years and decades. They are, after all, part of a physical process, the product of an interaction between two fluids, the ocean and the atmosphere. (To physicists the atmosphere is a fluid.)

Unfortunately, aesthetics or practical considerations notwithstanding, fluid dynamics, the branch of physics that governs the interaction between oceans and atmospheres, is fiendishly complicated. (An apocryphal story claims that both Einstein and Heisenberg took a break from relativity and quantum mechanics, respectively, to try their hands at fluid dynamics but gave up, claiming that the problems were just too hard.)

Nevertheless, three researchers at Columbia's Lamont-Doherty Geological Observatory were able to predict early in 1986 that by June 1986 the first signs of El Niño ought to appear. Led by Mark Cane, the team advanced the theoretical understanding of how El Niños happen by providing for Bjerknes's basic model a mechanism that could turn El Niño off. They reasoned that if the warm water from the western Pacific hit the Peruvian coast and then spread north and south, the pool of warm water would eventually dissipate to the point where the east-to-west flow of the trade winds could start up again and normal conditions could return. This also provided a hint as to why the cycle of El Niños is irregular, occurring every three to eight years. An El Niño can only recur when enough warm water flows back from the higher latitudes into the tropics, and that need not follow a precisely timed pattern.

That was the key: if the El Niño cycle depends on the motion of the surface layer of the ocean, then it ought to be possible to forecast the climate events by charting the behavior of that pool of warm water. Cane and his colleagues investigated the available data for the three El Niños that have occurred since 1970 to see if their theory would have predicted those events. They made six good predictions, at different periods before the event, for each El Niño.

However, the Columbia model's one prospective prediction for the June 1986 El Niño was wrong, sort of. The warming failed to show up on schedule. Starting in late 1986 and continuing into the spring of 1987, a classic, if mild, El Niño did occur, several months late. It's not clear what precisely went awry, whether the data the Columbia team was using was inadequate, its physical model in error, or its picture perhaps too simple to account for the range of variation possible when atmosphere and ocean interact.

And yet one doesn't want to give up on their idea. El Niños and analogous events are central components of the climate machine, important in the life of the planet and of the living things residing here. They are climate at its most obvious: periodic, dramatic changes in the weather across great stretches of the globe, lasting for months at a time. They are the most certain and easiest to grasp expressions of the idea of the earth as a system, integrated and complex. They ought to be understandable, as physical events; of necessity, we ought to try to anticipate them.

Of the two, understanding is easier than prediction. The description of El Niños and the other teleconnections is fairly complete at this point. Cane and his team used Bjerknes's basic theory, elaborate and refined, but still essentially the same after twenty-five years. Their model has survived several El Niños, but reliable forecasting is still elusive. Nevertheless, the Lamont-Doherty group is clearly on to something. Its success with past El Niños is genuinely impressive; perhaps the partial failure of its one true prediction was simply one of the inevitable glitches that come with any attempt to break new ground.

The effort to understand the underlying dynamics of El Niños takes climate science right to the boundary that divides randomness and some underlying order—between what patterns exist in climate systems and daily fluctuations of the weather. In this territory, under most circumstances very little can be foreseen; I cannot know, here in Boston, if the weather twelve months from now will be worse or better than it is today. But El Niños and the other teleconnections form a special case. They are not random; instead, they represent one of the few clear patterns into which the climate system falls, and as such they offer the best chance to determine why one year will be better than the next, or worse, much worse. For El Niño, there is still some absolute limit beyond which what the ocean-atmosphere system is doing today can tell nothing about what it may be doing some time hence. Cane and his colleagues believe that the limit is in the range of a year or more. Whatever the number, it defines a central property of the system that produces El Niños and similar events, the knowledge of which may provide us with some defense against the worst of the harm that they may cause.

Simple, true, and sufficient, but not the whole story. The

purpose behind anticipating El Niño is to allow ourselves to prepare for it, not prevent it, for that we cannot do. But the exploration of the links that tie the ocean and the atmosphere, one continent to the next, one year's conditions to the weather in the next has yielded yet one more novel vantage point from which to view our planet. Almost a century of labor has gone into uncovering the ways El Niño connects us to our habitat. But the notion of relation has been remade. Our misery has company from Madras to Lima.

There was drought in 1982, a killing drought. Elsewhere, the rains fell as if they never meant to stop. Seabirds died and seals abandoned their young to starve; people watched the weather turn enemy, and some of them lost their lives as well. Our world, our homes and streets, share a common bond created by a physical process that encompasses half an ocean and the atmosphere above it. El Niños, and by extension the year-to-year persistence of basic patterns of climate the planet over, have gathered us up.

CHAPTER **6**

Storm

THE STEAMER *Nan-shan,* once a registered British vessel, but now shipping under the flag of Siam, found herself one morning on the South China Sea under clear skies, bound for Fuchau with a load of coolies and mixed cargo. Her master, described as a man "having just enough imagination to carry him through each successive day, and no more," examined the barometer, saw that it was falling, and concluded "there must be some uncommonly dirty weather knocking about."

Joseph Conrad's story "Typhoon" begins calmly, quietly, with just the slightest indication that the corner of the world through which the *Nan-shan* steamed was about to be transformed by the sort of storm most of us will never see. Tropical storms, even hurricanes, typhoons are common enough; about eighty gales a year attain full storm strength, with about half to two-thirds of them going on to become hurricanes. But the sort of storm that struck the *Nan-shan* was a once-in-a-lifetime blow. Conrad, with his experience of the sea, tried to describe what such a wind is like:

It was something formidable and swift, like the sudden smashing of a Vial of Wrath. It seemed to explode all around the ship with an overpowering concussion and a rush of great waters, as if an immense dam had been blown up to windward. It destroyed at once the organised life of the ship by its shattering effect. In an instant the men lost touch of each other. This is the disintegrating power of a great wind. It isolates one from one's kind. An earthquake, a landslip, an avalanche, overtake a man incidentally, as it were—without passion. A furious gale attacks him like a personal enemy, tries to grasp his limbs, fastens upon his mind, seeks to rout the very spirit out of him.

However one experiences it, the weather is personal, as an enemy or friend. Its impact on human life is obvious and immediate. By tradition, at least, it is the weather that determines culture: New Englanders are tough and taciturn, hardened by the blizzards of winter; Californians, whose idea of weather is contained in the standard summer forecast "coastal fog extending inland night and morning, clearing by midday," lack the discipline imposed by the need regularly to batten down the hatches. Weather fits comfortably into the context of human life. The longest storms last for a few days and can be experienced as a single event; the great ones form the landmark around which we hang the memories of a particular time and place (there is no doubt that Dorothy must orient her recollections of childhood around that Kansas tornado, for example). Each storm is particular, discrete, a story with a beginning, a middle, and an end.

And every storm forms part of a story that extends far beyond the few days it takes for any storm to blow itself out. For climate science, the first task has been to draw an empirical picture of what it is that constitutes climate. On every time scale that picture has been modified. On the billion- and million-year time scales, climate is essentially a problem of geology and evolution, cataloguing what rocks, continents, and oceans exist; how they move and set up geochemical cycles; what living things exist, what they metabolize, and what waste products they produce. From hundreds of millennia to hundreds and thousands of years, climate appears as a dynamic system, a set of feedback loops that regulate conditions of temperature, atmospheric chemistry, and so on, within broad parameters. In the narrowest perspective, however, the issues become those of seasons and of weather, the mechanisms that control drought cycles or times of flood. Will it storm this winter, or will it be mild? At this point the picture shifts into focus.

A trick of perspective is involved: take the long view, and the weather is the noise, the chance twists and leaps, superimposed on the extended climate record; look in the other direction, starting from days and moving through years, decades, and millennia, and climate is an accumulation, the sum of all the individual storms and spells of fine weather. Each individual storm has its own life and death; but storms recur, of course, over years, decades, millennia, and it is to storminess, not a particular storm, that the living things in the path of the weather adapt. At one time naturalists looking through early microscopes thought they could see the whole world re-created in miniature in a drop of water. A single storm gives something of the same experience: its details are its own, but it provides a glimpse of the web of connection between climate and life that has evolved over all the billions of years. A storm is a metaphor for storminess, so by examining closely what is unique to a particular storm and what features of it are common to all such events, we can begin to define precisely the ways in which the chances of the weather become the constants of climate.

On September 4, 1938, weather instruments at the Bilma oasis in the south-central Sahara Desert registered a weak atmospheric low. The disturbance was mild but picked up some strength as it moved west. By the time it reached French West Africa on the seventh, the collision of the low with the established wind systems—the trades, the southwest monsoon, and the equatorial easterlies—set up a whirling motion. The wind formation continued to move west and settled into a cyclone's circulation pattern over the Atlantic by September 13. By the morning of the sixteenth it had reached full hurricane strength (which is defined as a storm with winds greater than 75 miles an hour). The storm traveled across the Atlantic on a roughly east-west track until early on the nineteenth, when it turned toward the northwest, passing above Puerto Rico, Haiti, and Cuba, and then began to run north, parallel to the American coast. By 7:30 A.M. on September 21, the storm center lay 120 miles off Cape Hatteras on North Carolina's outer banks. Remaining over water, the hurricane continued up the coast.

Under ordinary circumstances, this storm ought to have bent to the east and eventually died out somewhere in the North Atlantic. However, for the past several days a mass of polar air had lain against a mass of warm, tropical air in the atmosphere over New England, forming a front that by the twenty-first lay directly above the Connecticut valley. The strong temperature difference between the two masses of air set up a powerful local wind system, with the air currents flowing from the south at from 40 to 60 miles an hour at a height of 10,000 feet. The winds acted like a funnel, sucking the hurricane onto and over the shore. The storm made its landfall on Long Island, crossed the sound at New Haven, and then cut a swathe through central New England. By the time it finally petered out, just to the north of the Great Lakes, it had become the worst hurricane to strike the region since English colonists first recorded the weather.

Even after fifty years, people who lived through it can still reconstruct what they were doing hour by hour, almost minute by minute, during the day that the hurricane struck land. A man I know was walking along a Rhode Island beach recently when he came upon an elderly woman staring straight out over the Atlantic. When he asked her what she was doing, she told him that she had come back to look at where she had lived, back before the storm of 1938. Edward Lorenz, now one of the deans of American climate research, had just taken an entrance exam for Harvard's graduate school that week and was set to drive up from Cambridge to Hanover, New Hampshire, on the afternoon of the twenty-first. The storm seemed to dog him, very much the personal force that Conrad knew, blowing him north, leading him up one road, blocking him there with floods, forcing him back into Bristol, and then trapping him in town with fallen logs across every exit route. Lorenz can rattle off each mishap along the old New England roads.

The statistics are more prosaic, of course, but they also tell the tale. Something like 600 people died, though no one ever completed a precise tally. Nearly 9,000 homes and summer cottages were destroyed outright, along with more than 10,000 barns and other buildings. Some of the 2,065 boats that were smashed or sunk came to a strange end: the railroad link between Boston and New York was cut for about a fortnight while an emergency work crew of 10,000 men repaired the washouts, rebuilt the bridges, and removed obstacles from the line—including a steamboat of considerable size that remained athwart the tracks for seventeen days. And worst hit of all, probably, were the forests. Five years after the storm, the U.S. Forest Service published its account of the storm's toll: the winds and floods associated with the hurricane devastated 500,000 acres of woodland, almost completely destroying the timber stands there, while causing at least some damage over another 15 million acres—one-third of the total land area in New

England. In all, an estimated *2.65 billion* board feet of timber blew down in a day and a half.

It was a unique event, still regarded as the greatest single natural disaster ever known to have struck the Northeast. By itself, the hurricane illustrates the extraordinary combination of violence and caprice that the natural world can bring to bear on the world created by human beings. It was the kind of storm, the kind of day, for which there is no preparation.

By chance, *In Hazard*, by the British writer Richard Hughes, was published in 1938. Based on the experiences of an actual ship, it was an account, modeled on Conrad's tale, of a storm of incredible power that in fact did batter the real ship to the very edge of destruction. Hughes had a sense of the inequity inherent in nature and noted the fact that the storm laid bare the flaws in men and objects that would have remained hidden forever but for that chance encounter with the wind. In Hughes's story, the ship's captain afterward tried to find a pretext to avoid mention of one crew member's cowardice in his report:

He had broken down in the storm, that was true. But that was bad luck. A man has no right to have to face such a storm as that. He might have gone through his whole career without ever a fault if he had not been so unlucky as to have to face that storm. The odds were heavily against his ever having, in his career, to face such a storm again. . . . An efficient officer: broken through one piece of bad luck.

Bad luck indeed. Hurricanes are not ordinary events for New England; they don't strike regularly the way they do in the Caribbean or along the southern and southeastern coasts of North America. But storms of the magnitude of the 1938 hurricane are extraordinary by any measure. After that hurricane, a young Yale-trained forester named David Martyn Smith compiled a catalogue of severe storms in New England. He found that storms strong enough to do significant damage come along on average at least once a decade and that even the monster hurricanes recur. On the 14th of August 1635 a storm passed close enough to the Colonies to do enormous damage. "It blew down many hundreded thowsands of trees, truning up the stronger by the roots, and breaking the higher pine trees off in the midle, and ye tall yonge oaks & walnut trees of good biggness would like withe," Governor Bradford of the Plymouth Colony wrote. In 1815 another enormous hurricane crossed central New England; aside from the 1938 storm, it was probably the most severe hurricane to strike the region since colonial times. Salt spray blew forty miles inland (as it did in 1938) and according to one nineteenth-century account,

The loss of timber trees was exceedingly great; and in order to save as much as they possibly could from the ruins of their forests the owners had the logs sawed into lumber, with which they constructed houses, barns and other buildings. Probably New England never knew another season of such building activity as prevailed in 1817 and 1818.

From these accounts, it appears that killer storms strike the Northeast about every century and a half. Three storms don't make a trend, of course, and in any event, that conclusion is a statistical statement: disastrous storms could just as well strike in successive months, with truly bad luck for those involved. Over enough time, however, the frequency of such extreme events probably evens out so that no one who survived the blow of 1938 is likely to endure anything comparable in his or her lifetime.

Play with perspective again. The storm was unique to its time and place. As one of a series of storms, too greatly separated in time to appear as a series in human memory, the hurricane of 1938 offers a chance to see how the events of days affect processes that take deades or centuries to unfold. From fifty years on, though, its impact on the landscape, if not on people, begins to fit into a pattern. Walk today in the New England forest and look in the right directions, and you can still see traces of the storm's passing, working themselves out in the regeneration of the woods.

The original New England forest, at least in the central Massachusetts area and for some miles north, was a mixed hardwood forest interspersed with a little bit of pine. When the first settlers reached the Berkshires, they cut their farms out of oaks and maples. By the middle of the nineteenth century the farmers were numerous enough to have cut down vast stretches of the original timber; in fact, one hundred years ago rural New England had far less forest cover than exists today, along with more miles of road inking farm to farm, than the region can boast now. The bottom fell out of the hardscrabble agriculture within a very few years, however. The railroads, in particular, which began carrying food from the Ohio valley and further west, drove the New England farmers out of business. Country boys moved into the cities or out West, and the fields lay fallow. Then, in the mass of abandoned open land, a forest began to reappear, not hardwoods this time, but white pine by the thousand, forming a resource that supported a variety of manufacturing enterprises. Small mill towns formed, with factories producing barrels, matchsticks, toys, and other soft-wood products. To support the new industries, foresters attempted to cultivate the pine woods even at the expense of other species. By 1938, hardwoods had begun to recapture some of their range, but in the main the pines still dominated.

The hurricane changed all that. What the storm did was to force the forest to start over again, from the beginning. The U.S. Forest Service did its best to salvage the downed trees, running logs into hundreds of ponds turned lumber pools for the emergency, and the new woods were able to form an open ground, in bright sunlight. The pines were the first back, just as they had sprung up in the open fields, but this time they were quickly shouldered aside. Researchers at the Harvard Forest in the hills around Petersham, Massachusetts, kept a running tally of the returning trees, and within ten years after the storm their land was dominated by a mixture of species—pin cherry, red maple, red oak, and white ash. Forty years after the storm, the tallest, largest trees in the forest were the red oaks and paper birches, with maples and white pines prospering as well, particularly in areas where they did not compete with the larger hardwoods.

Today in the Harvard Forest you can still come across huge blow-downs, uprooted trees the Forest Service missed. The forest that has grown up around these relics is a new wood, of course—almost all the trees are younger than the oldest scientists who now study them. What's striking, though, is that this forest is, in one sense at least, far older than the age of any individual tree. The forest that blew down in 1938 was in some sense an artificial wood created by the clearing of the native forest for farms and the cultivation of the pines for industry. The succession of species since the hurricane sped up the changeover from the man-made forest into the kind of woods that would occur in a New England without people; the forest today now looks at least somewhat like that that the English would have seen when they landed, a balance of species that has been absent from much of the region since before the Civil War. The storm, which seemed a disaster, an absolutely extraordinary event, becomes in the Harvard Forest an agent of historical continuity that binds the fifty-year-old forest standing now to the forest of the precolonial era. Within the forest, the two days of the storm lie embedded in a process that extends across centuries.

A kind of forecasting is implied in the response of the forest to the storm. The forests of New England have evolved into an ecosystem that expects periodic major disruption, and they retain the means—stores of seeds, a mix of species that can recolonize cleared ground, and so on—to deal with the problem when it arises. This response is a kind of prediction based merely on the fact that storms recur from time to time, but not on the need to gauge precisely when a storm will strike. Such knowledge, of course, is scant consolation for anyone who lost property, friends, or family back in 1938 and scant help to anyone worried about when the next storm is going to strike his new house on the beach, his family, or his own life.

That is what prompts society to demand of those concerned with the behavior of nature that they should be able to figure out how to keep the natural world from mucking up the affairs of people. Scientists ought to be able to keep nature at bay, or at least to provide adequate warning when they cannot. It continues to come as a surprise when the weather disrupts modern urban life. In 1977, a dean at Harvard swore that the university would never close "except for an act of God, like the end of the world, maybe." A year later, the blizzard of 1978 shut the place down for a week.

A healthy storm always has the power to make the world we live in seem more vulnerable, more precarious than usual. In the autumn of 1985 New England was again struck by a major storm, Hurricane Gloria. Gloria traveled along almost the same track as did the hurricane of 1938. Throughout the day on which it passed near Boston, the city's network television affiliates devoted their entire airtime to covering the storm. Many of the broadcasts consisted only of announcements about emergency measures or pictures of the storm's impact further south, but about every fifteen minutes or so each station cut to its weatherman for any prediction, any clue about the hurricane's behavior that the forecaster had missed a quarter of an hour before. A viewer could not help but be impressed by the apparently endless panoply of tools and props each of the weathermen brought to bear on Gloria. There were the measuring devices—barometers, wind speed meters, rain and tide gauges—as well as more sophisticated interpretive equipment like radar, satellite imagery, and computers to manipulate the data.

Throughout the day, Gloria blew and the weathermen pointed, predicted, and described; yet despite their impressive tools and collective expertise, they were unable to answer some of the most pressing questions that the storm posed. They couldn't say when it would strike the New England coast, or how strong it would be when it arrived. The storm overtook them, striking where it would at an intensity that surprised and then passed on. Before the storm arrived, the weathermen couldn't say terribly well what she'd be like; once she was here, the wind and the rain told their own story, and nothing a weatherman could add could tell it better. Like the rest of us, the weathermen simply had to wait and see.

Hurricanes are extreme examples, and it is not unreasonable to argue that they could present intractable problems for the forecaster. They do, but even as special cases, they mark out the limits of what actually can be said about the weather. The hurricane of 1938 is an example of the uncertainty bred into the science of weather; in its violence and the freakishness of the atmospheric conditions that drove it forward, that hurricane emphasizes the singular character of each great storm, that aspect of the weather that still can confound the forecaster. But hurricanes as a class of storms do share features, and those characteristic patterns give the scientist the clue with which to learn how much of the weather he can, in fact, predict.

In some ways, hurricanes are very simple storms. They all share the same fundamental structure, and they all form within a consistent and quite precise set of conditions. Tropical cyclones, which above certain wind speeds may be called hurricanes, always form over a sea surface whose water temperature is at least 80°F. (Plenty of storms form over colder water, but they don't spin like cyclones, and they don't become hurricanes; what magic is in the 80°F threshold, no one knows.)

Over warm water, an atmospheric disturbance like the low observed in the Sahara early in September 1938 creates a flow of air upward, with warm, water vapor–laden air rising from the sea surface, feeding the formation of clouds and perhaps of thunderstorms. This flow of air creates a low-pressure area at the sea surface, and air from outside of the disturbance begins to flow inward, following the gradient from higher pressure to lower. If all this occurs right on the equator or within just a few degrees of it, what you get is a nasty storm, not a hurricane. But if the storm develops more than 250 or 300 miles away from the equator, the spin of the earth around its axis will bend the winds as they flow into the center of the vortex, and quite quickly the whole system will begin to rotate.

As the storm intensifies, it relies on a network of feedback systems to keep it going. The warmer the air temperature, the faster the circulation of air into the bottom of the storm and up and out the top will be. The faster the storm travels, the more moisture it can carry. That in turn drives the winds still faster because the rain falling from the heights of the clouds releases heat carried up from the ocean, which in turn enables warm air from the surface to flow upward with still more vigor. (That, in fact, is why hurricanes cannot form over land and die out quickly as they drag across a continent: there is no water to keep the storm fueled.)

Simple, after a fashion. Unlike other storms, hurricanes can exist only under certain fairly well-defined circumstances; when those conditions exist, we know that we are at risk. Chance still plays a major role; it is not clear why some storms develop into hurricanes while others do not, even when the situation seems ripe for a big blow. The transition from a powerful tropical storm to a hurricane can occur very rapidly, surprisingly so at times. But at least when a hurricane does form, we know essentially how it will behave. It will travel, carried along by the general circulation of winds within the atmosphere, picking up moisture and dropping rain until its supply of heat and water vapor gets cut off when it is either over land or over oceans that are too cold to sustain it.

But there is a limit to what we can say about the specific development of any individual storm before it happens. The hurricane of 1938 followed a path that is perfectly intelligible, in retrospect. It formed in the ordinary way; it traveled as such storms customarily do; it picked up moisture and energy in the tropics as usual and then prepared to expend them in the higher latitudes; then finally it encountered the freak atmospheric conditions that sent it catapulting across New England. And from that freak of the weather flowed all the tallies of destruction and hurt. After the fact, the physical processes that made the 1938 storm a special agent of disaster are easy enough to discern: the combination of the high tide when the storm made landfall, the front up the Connecticut valley, and the unusual strength of the storm itself. At the time of the actual event, though, the crucial details about where the storm would land, what route it would travel, and how much damage it could do necessarily came as a surprise.

The details had to, because the climate system is one that has a historical memory but not a dynamical one. Great events leave their traces. Climate persists, leaving its record in ice and mud, in the rings of trees and layers of rocks; in this sense the earth has a record of what has happened here over periods ranging from seasons to billions of years. From this record with a kind of recollection you can make certain kinds of predictions—the sort that come from recognizing that a forest in New England is accustomed to storms of a certain magnitude. With that kind of knowledge and the discovery of the processes that drive events like the ice ages, we can make a stab at predicting how stormy the weather will be over time, or what kind of temperature fluctuations to expect from decade to decade or millennium to millennium.

Unlike the earth as a whole, however, the atmosphere forgets quite quickly. Patterns recur—El Niños or the regular push of the trade winds—because they result from the interaction of parts of the system that remain constant throughout the changes, but even within El Niño conditions, the particular details of the weather on one day will still allow no clue as to the state of the weather two or three weeks hence. The broad El Niño state will obtain, but it is unknowable whether the wind speed over a given point will be five knots or twenty, and so on through all the relevant meteorological parameters. The atmosphere lacks any recollection of the precise combination of winds and clouds and pressures that obtained even a short time ago. It is as impossible to recall the past as to predict the future with knowledge only of the atmosphere's present. Looking out the window as I write this, I see two layers of clouds and a scrap of blue sky; the clouds have been blowing in from the west since morning, and the weather now looks much as it did three or four hours ago. But nothing I can find by looking at the sky or by examining the weather map for today can tell me what the weather was doing a week ago: the conditions within the atmosphere change so rapidly that the sky today has forgotten what it was like those few days ago.

The root cause of this unpredictability is the process of *advection*, the wind's carrying of such properties as heat and cold or dryness and moisture. Winds behave in a nonlinear way, so that when one wind system interacts with another, the result is not necessarily the sum of the two movements and in fact may be a situation quite unlike either of the two winds considered by itself. As different systems of winds interact in ever less predictable ways, what the atmosphere will do next becomes increasingly hard to determine. In 1956, Philip Thompson, then with the Air Force, made the first estimate of the time limit beyond which forecasts become guesses: he calculated that after two weeks prediction would be impossible. There has been progress: Thompson, now with the National Center for Atmospheric Research (NCAR), has since figured that, in theory, the outer limit of weather prediction for an area the size of the northeastern United States is perhaps three weeks. Beyond that the great range of possible changes open to an atmosphere starting from a given set of conditions just becomes too great to do better than guess.

Hurricanes pose a special kind of problem for prediction. Unlike most weather systems, they are not subject to certain kinds of change. They fall into a class of objects called coherent systems, that is, anything smaller than the storm itself doesn't affect the structure of the storm; for instance, a thunderhead encountered on the way will not alter the basic shape of the hurricane. Because every hurricane always behaves like a hurricane, it's possible to make the broad prediction that a certain number of hurricanes will hit the Northeast every century. But there is a catch, of course: coherent systems can be manipulated by elements larger than they are. For hurricanes, the larger force that matters is the general circulation of the atmosphere; the vagaries of the atmosphere as a whole determine where and when the hurricane goes. So what you would most want to know—when a storm is going to hit, and if it is going to hit your particular turf—is exactly what is impossible to determine more than a very short while in advance.

MIT's Edward Lorenz, with Thompson one of the founders of predictability research, has an elegant way of expressing the predicament: Human beings cannot attain perfect knowledge, but what keeps us from knowing what will happen to us next is our inability to describe exactly what is happening at any particular point in the present. Even a very slight gap between our image of the weather and the events as they are actually occurring, right now, in nature, will generate errors as we try to determine the future. Over days and weeks the errors multiply; the gap between what we expect and what does happen expands; and then the future grows dark. In what has become one of the clichés of the atmospheric sciences, Lorenz asserted that the flutter of a butterfly's wings is perturbation enough to transform the whole system, over time. More prosaically, Lorenz compares the atmosphere to an old-fashioned pinball game, the kind with balls and pegs, but no flippers or springs. Imagine a tiny difference between the angle at which you think the ball bounced off the first peg and the angle it actually took, perhaps deflected by a single scrap of dust in the machine. At the next peg the path you predict and the path the ball travels will diverge a little farther, and the error will grow at the next peg, and the next. You can follow the ball for a bounce or two, but by the time it hits the third or fourth peg, its course is entirely different from what you expected after the first collision.

It is a rather grand leap from the *Nan-shan* to an errant pinball, from the personal relationship between men and their corner of the natural world to the anonymous work of varying winds, of specks of dust, of all the interacting minutiae. Part of the jump derives from the contrast in times: Conrad is a nineteenth-century man and Lorenz belongs firmly to the latter part of the twentieth. For Conrad, even the unfathomable enmity of the wind could be countered by the order men could create, such men as his captain, who possessed just the required amount of imagination. The notion raised by Lorenz's work, the idea that there is some absolute limit to certainty, is one of the cornerstones of twentieth-century thinking. At the heart of science is the act of measurement, and although the absence of measurements sometimes appeals to working scientists (without data to constrain them, they can let their imaginations run absolutely wild), fundamentally, if you can't weigh it, count it, time it, or somehow tie a number to some aspect of it, you can't prove or disprove any broad statement you might want to make about the nature of what you're looking at. It could be beautiful, but it wouldn't be science.

This reverence for measurements is an uncomplicated creed, and a powerful one. It forms the pragmatic core of science's claim to be the arbiter of what is real: a good, solid appreciation of the ability of facts to counter flummery. At the extremes, this attitude can lead to its own errors: witness Lord Kelvin's insistence that the age of the earth, based on how much cooler the world is today than it must have been when it formed, was only ten million years—a "fact" that lead him to conclude that far too little time had passed for Darwinian evolution to occur. Lord Kelvin was right that ten million years was too brief a time for natural selection to operate, but, being ignorant of the heat given off by radioactive elements, he was wrong on the planet's dates. The situation now is different from that of Kelvin: what Lorenz showed was not that measurements were in error, but that the most accurate measurements could not yield accurate results, and what Lorenz demonstrated for meteorology has since proved true of a number of sciences. There exist some absolute limits to what human beings can know about their surroundings.

From this heart of uncertainty derives the final power of a great wind: it retains always a core of mystery. Richard Hughes wrote that he was called in by the shipowners whose vessel had actually weathered the storm that rises within *In Hazard* because they "felt that an event so extraordinary must never be forgotten. They had sent for me as for some kind of tribal bard." Hughes himself saw a connection between the madness of the elements and the disintegration of Europe that proceeded as he completed his story. The uncertainty, and the inability to prepare for or prevent the worst that the storm could do became for him a powerful metaphor for the broader events of his time. I don't want to press the issue too far, but I believe that an increasing awareness that there are limits placed on what human reason can accomplish has contributed to the precarious feel of the current day, the sense that things may go badly wrong in a way no amount of clever thinking can prevent.

What the bard or the scientist can do, however, is to ensure the continuity of historical memory: this function goes a great deal of the way to minimizing the consequences of uncertainty. To paraphrase an overused saying, climate begins with a single storm. Events on one time scale can mirror events on another. A storm becomes a paradigm of storminess; frost a microcosm of the decades of a little ice age; the swing from winter to summer forms an analogue to the transition from glacial age to interglacial. And in the line of metaphor, the unpredictability of the weather becomes the inevitability of climate change. The system, climate and weather both, once again demands a continuous flexibility of perspective, of how you choose to look at the world. Uncertainty over a brief time becomes a kind of certainty as weather stretches into climate. You can be almost completely sure that the characteristic weather in the place you live will be different fifty years from now than it is on the day you read this.

Living systems—like the New England forest, or a community of wetland grasses, or the desert—have come to terms with climate changes and the unpredictability of the weather. By all the evidence, human societies rarely do as well. For just one example out of thousands, the beaches along the eastern seaboard of the United States are more built-up now than they ever have been, with most of the construction having been done during a period when no truly devastating hurricanes blew through to remind people of how fragile a beach cottage can be. As a result, the next hurricane comparable to the 1938 storm will have a far greater potential for devastating harm than any that came before it. There are alternatives to such foolishness, but no general answer, no way to avoid all risk. It is clear that people should not build on the water's edge, for example, but as the memory of the last great storm has faded, the sense of risk to life and property has disappeared. Good science can help reinstill caution, but it takes a Conrad, too: a good story, a grand storm, can lodge firmly in one's mind.

And from a storm, a single bout of bad weather, into history: from a day, a year, decades to centuries, millennia, hundreds of thousands of years, millions, billions. Climate evolves over every period, from a single day to the planet's lifetime, and as weather it is evolving still, connected with and resting upon events that transpire over every time scale. Oxygen is still being produced and regulated (thankfully); the rock cycle still processes carbon from land through air and ocean and into the deep recesses of the earth. The planet continues its uneven way around the sun, and plants in the ocean continue to process carbon dioxide themselves. It is cooler now than it was in 1066; the El Niño of 1986–87 has receded, and the weather in Boston tomorrow, as I write this, is supposed to be a little warmer and a little damper than it has been today. This historical perspective forms the first great accomplishment of the new field: the creation of a vantage point from which the world and its entire history appear whole, upon which events that occur have antecedents stretching out over both space—the entire globe, often—and time, even time reckoned in billions of years.

PART II

ELECTRONIC WINDS

A CHANGE in the observed demands changes in the observer. Much of the effort that has gone into creating the new science of climate, the science of this integrated whole, has been work of a traditional, familiar sort: digging up the data piece by piece and then jimmying each one into the theories that attempt to describe the system as a whole. That approach has served to explain a great deal of the specific phenomena: ice ages, hurricanes, and El Niño. The presence of oxygen in the atmosphere and the existence of a fairly narrow band of temperature ranges throughout the history of the earth have all yielded at least some of their secrets to this kind of approach.

But now the problem has become understanding what drives the continuing evolution of this system-of-systems—the oceans, land, atmosphere, biosphere, ourselves—and the bonds that link the entire assemblage together. To do so requires new methods, a new practice of science, and, ultimately, new scientists, for it is now more or less completely impossible for a climate researcher to run a traditional experiment. There is no way that a laboratory setup can match the complexity of the whole climate or that an individual investigator in the field can change one of the climate's constituents, hold everything else constant, and measure the results.

The leading characters of these next three chapters, then, are scientists who have come up with an alternative to the ordinary drudgery of fact gathering and the lab bench. But the hero or *éminence grise*, depending on one's outlook, isn't any one of the researchers; it is, rather, the machine they have come to use as a stand-in for the irreducible complexities of the real world: the computer.

Revolution is a hackneyed word, particularly in science, but in this context I believe it appropriate. Climate science lives and dies by the computer; it cannot exist without it. In the old days the scientist looked at nature and back again to his own thoughts. Since the Second World War in many sciences, but especially in this one, the relationship has become a triangle, between the observer, the natural world, and the simulacrum of that world that the scientist builds inside the machine. What follows, then, is a romance: a tale of revolution and of the tangled affairs of those who fought its battles.

CHAPTER 7

When They Were Very Young

S OME SCIENTISTS are brilliant, and there is no accounting for their genius—where it came from, why it is that one person can see more deeply than his colleagues. Some scientists simply have the "knack." At first glance their success looks like luck—they asked one right question, they chose one good, uncluttered field of study, something just happened right. But over time with a number of these individuals it becomes clear that something other than luck intervenes: they consistently ask the questions that allow them to find interesting answers; they know where to look to get the most information with the least effort; they have a sense of where the cutting edge lies.

When a new science, a new line of inquiry, forms, the logic that impels it forward is obvious only after the fact. Climate science was born of a synthesis of a variety of fields that had, and still have, a great deal of uncharted territory. The links that formed between the variety of disciplines that now make up climate science— physics, chemistry, meteorology, geology, and so on—did not emerge piece by piece out of work undertaken to solve the particular problems of each of those individual sciences. They were forged, rather, by a group of people who, trained within one area of discourse and pursuing one particular problem within that area, found themselves able by accidents of timing, temperament, and skill to reach past the fetters of a single intellectual tradition and peer into the wider realm that the concept of climate exposed. They were, in fact, scientists with the knack.

In the summer of 1940, Walter Orr Roberts set out from Boston and headed west toward the Rocky Mountains. He had left graduate school at Harvard for a minehead at the Continental Divide, where he intended to create an observatory dedicated to studying the physics that govern the sun, but during his journey another question intervened.

Roberts was then a newly married man, and when Harvard gave him a train ticket west, he cashed it in to buy a quite used Graham Paige automobile that could transport him and his bride of one month. It broke down for the first time in Wellesley, Massachusetts. Crossing the Midwest in the summertime, the couple could drive only at night because the Graham Paige overheated almost at the mere sight of the sun. This left Roberts with time enough to look closely at the countryside they reached after each night's drive. To Roberts, who had spent his boyhood on and around farms in New England, the midwestern dust bowl four years after the end of its great drought seemed a blasted land:

You have to remember that the dust blew, the winds were strong, and they blew all the way across the west from the Rockies practically to the East Coast. The dust was picked up and carried to a height of 10,000 feet, and when it rained in Kansas City it rained mud. The dirt from the fields drifted like snow, and piled up behind the buildings; the people had to leave. You would go into some towns, little Nebraska towns that had a central park, sort of, a grassy area with trees around it. The trees were all dead, and it was just a miserable scene. You could see the people who stayed behind—the whole vitality of the town was gone and we went through town after town like that. I had never seen anything like that.

For the next five years Roberts occupied himself with his original study at his observatory in a settlement in the Rockies called Climax, Colorado, 11,000 feet up. The Climax Mining Company was the town, really, and it gave him a house that stood dead on the Continental Divide. The rain that fell on the east eave of his roof drained into the Atlantic Ocean; the western gutterspout fed the Pacific. Roberts discovered early on that there was a correlation between solar activity and whether radio communications could get through interference from a solar storm. For the duration of the Second World War, Roberts remained there for as long as eighteen months at a stretch, making communications forecasts for the military, but finally, with the peace, he was able to turn his attention back to the question of what could cause the devastation he had seen in the dust bowl.

When Roberts first began to think about the weather, the science of climate was barely twitching. Weather forecasting of a sort did exist, but not a concept of the system operating as a whole. That idea developed incredibly quickly though—in the five years after the end of the war, perhaps, or in the couple of decades running from the forties to the sixties, or in the working life of a scientist whose career stretched from the golden age of physics in the thirties through to the present day. Roberts did not create the field of climate science—no one scientist can claim more than a piece of it. But he is one of a handful of men who nurtured it, who asked the crucial questions that defined it.

The first step was the one implied in Roberts's intuition about the connection between the daily experience of the dry days that created the dust bowl and the quirks of solar physics that might have an impact on climate. Generalized for a range of problems, the strategy was to figure out the weather first and then to worry about the larger system of air, ocean, life, and land, and about the time horizons that extend beyond events of a day or two. As a matter of history, climate science emerged out of meteorology. But to create a sophisticated picture of the weather and then to build from that a scientific approach that could encompass the evolution of climate, the central concern of meteorology had to shift. The practical problem of what the weather would be like tomorrow or for the next planting season needed to be recast as the more abstract issue of defining the underlying mechanisms that could create a difference in conditions hour to hour, day to day.

At the beginning of the Second World War, however, the central concern was to generate forecasts upon which pilots, soldiers, and sailors could depend. Meteorologists, including those upon whom the military relied, had little underlying theory to guide them or to give them any confidence in the detailed accuracy of their conclusions. The British had developed an extraordinarily sophisticated network of weather stations during the First World War, with which they were able to issue moderately reliable predictions. They did so by drawing up weather maps, based on observations of atmospheric pressure, wind speed, temperature, and so on, at as many different sites as they could manage. They would then compare those maps with maps of other days until they found a close match in conditions; at that point they would see what weather had followed the earlier case, then predict that more or less the same events would happen in the present. Weather forecasters were thus essentially well-organized weather historians, a shade better equipped than a down-east farmer, perhaps, but lacking any real methodological advantage over someone with a long and detailed memory.

Nonetheless, this wasn't that bad an approach, but it was one that was absolutely limited in its accuracy. Such comparisons depend on two assumptions: that the atmosphere will behave a second time in the same way it did earlier; and that you have enough information from enough sites so that you can be reasonably confident about what the atmosphere is doing. Both of these assumptions fare poorly, the more so as you get further and further down the road from the moment the readings were taken. In 1922, according to climate theorist Stephen Schneider, the longest range of a forecast for the British Isles was three days.

In the face of such difficulties, particularly under the pressure of war, certain people turned out to have what seemed to be a special "weather sense" that allowed them to tease predictions out of unreliable data. One in particular, the Canadian forecaster Patrick McTaggart-Cowan, was famous for his feel for the weather over the Atlantic, and supposedly some pilots who ferried planes across to Europe would not cross the water without a McTaggart-Cowan forecast.

It was unacceptable in wartime, however, to rely on some kind of occult faith in the gifts of weather seers. The demands of modern air combat and of massive amphibious operations provided the impetus to break the mold. The crucial conceptual step came when several scientists tried to apply fundamental laws of physics—the laws that govern motion, heat transfer, and the behavior of fluids and gases—to the atmosphere. Actually, the war accelerated a program already under way. A very few men had attempted in the two generations that preceded the war to piece together the structure of the atmosphere. Gilbert Walker in India, with his measurements of the shift in barometric pressure over the Pacific, was one; the underlying pattern of winds is known to this day as Walker circulation. In the 1920s a Norwegian father-and-son team, Vilhelm and Jacob Bjerknes—the same Jacob who finally completed what Walker had begun, discovering the mechanism that linked India's monsoon with Peru's El Niño—devised a fairly simple theory of fronts and cyclones that explained the creation of storms well enough to permit forecasts several days ahead. Crucially, these were genuine predictions, derived from an understanding of the motion of air within the atmosphere and not simple correlations between conditions on two separate days.

Traces of this kind of inquiry appear even further back in history. Perhaps the first practitioner of climate science was the Englishman George Hadley, who was in charge of meteorological observations for the Royal Society. Noting that different parts of the globe received different amounts of energy from the sun, he proposed in 1735 a model of planetwide atmospheric circulation, with warm air from the equator entrained in a current that flows to the poles and back again. Unfortunately, his theory was not borne out by reality. The rotation of the earth does bend the currents of air, as Hadley proposed, but produces instead great rings of circulating winds, which are nonetheless now known as Hadley cells.

Hadley's idea appears remarkably prescient, as it was; but from Hadley forward, pieces of the puzzle accumulated, without the unifying pattern. It was still impossible, without very particular conditions like those noted by the Bjerknes team, to predict the weather based on an understanding of the processes that govern events within the atmosphere.

Walter Orr Roberts's first crack at the attempt to discover a fundamental unifying theory of the weather was, at least partly, a failure. Released from the demands of war work, Roberts returned to the question that had originally struck him on his drive across the dust bowl back in 1940. His first line of attack was to seek a connection between solar activity and large-scale climate patterns. While still in graduate school, Roberts had first encountered the idea of such a connection in the work of Charles Greeley Abbot, but he had not given it much thought at the time. Abbot, an astronomer who studied the eleven-year cycle of sunspots, argued that droughts, or dust bowls, in the Midwest occurred about every twenty to twenty-two years and that the dry spells were somehow linked to a process that occurred with every other sunspot cycle.

Abbot predicted that the next drought would hit the Midwest some time in the 1950s, as it did, and subsequent research has established at least some evidence for a twenty-two-year drought cycle back through several thousand years. The link Abbot identified is a statistical one—the physical mechanism by which sunspot cycles can influence the weather is still unknown—but the correlation does exist. Roberts reasoned that if sunspots could influence the climate every twenty-two years, solar activity of some sort ought to have some bearing on weather and climate on a much shorter time scale.

So Roberts set out to teach himself about the weather. He returned to Harvard as a professor for the 1947–48 academic year. He read; he sought out some professional meteorologists; he had lunch three or four times a week with one of them to talk about the sun and the weather; and then he returned to Colorado to observe the sun until he could establish some relationship between events there and events here. And he established—almost nothing. Roberts and some colleagues did discover that in the Gulf of Alaska barometric pressure would change after a solar event that triggers a magnetic storm around the earth. It is a mild effect, it does not help weather forecasting in any way, and it seemed to be a purely local event, confined to a small portion of the northern Pacific. Even worse, from Roberts's perspective, was its similarity to Abbot's twenty-two-year cycle: even though the effect seems to be real, the evidence for it is purely statistical. Roberts to this day hasn't a clue as to how it works.

In essence Roberts gambled, and lost, and won. The gamble was to assume that there is a day-to-day solar-weather connection; the loss was in the years of effort trying to ferret it out. The victory was a little more elusive; it came, though Roberts may not have known it at the time, because there were actually two sets of stakes on the table. The first was the conventional prize of science: the chance of a discovery, of a breakthrough, which Roberts missed.

But his question was the right one, or among the best of the right ones to ask then. It was an *integrative* question: How do fluctuations in the transmission of energy into the climate system affect the weather?

What Roberts won in the end, in concert with a handful of other workers, was the chance to create a new field, to play in the wide-open spaces that quite suddenly came into view in the few years just after the war.

Other scientists were then pursuing similar notions. One of them, Herbert Riehl at the University of Chicago, took advantage of the practical research that had been done on weather in the tropics during the war to investigate the impact of tropical mete-orology on weather in areas far removed from the equator, beginning, with a number of other scientists, the effort to tie together the ocean and the atmosphere. By the early 1950s, a line of research that had stretched back to George Hadley, back to the eighteenth century if not before, had finally produced a picture of how the planet functioned that reduced, although it did not eliminate, the sense of accident, of unpredictability. From the prospect of weather prediction based on an understanding of atmospheric dynamics it became possible to conceive of a new enterprise, one that binds this newly understood (more or less) arena of the atmosphere with the rest of the planet, and in time, with the history of the planet.

After his initial attempts to construct climate theory, Roberts spent the bulk of his scientific career organizing the context in which productive research into climate could take place. What distinguishes Roberts most of all is his timing: he set his sights on a new science at the first moment that it became possible to conceive of ways to find answers to the questions that a great number of extremely bright investigators had previously posed and had previously been unable to pursue. A science requires more than an interesting, valid idea to drive it forward; theory, specu-lation requires experiment, some material connection to the mate-rial world of nature with which it is concerned. Up until this period forty years ago, essential questions of climate science were untest-able, immune to experiment, and hence the science could not develop. It was impossible then to conceive of the classical kind of scientific experiment—setting up certain conditions, altering one, and observing the result—when the subject was the entire world; it is impossible now. What ended the methodological impasse was the creation of a tool that built, in effect, a novel window from which to peer out of theory and into the weather.

There stand memorialized in the lobby of the computer science building at Harvard the bones of a dinosaur. At one end stands a bank of counters, each more than twenty digits long, each of which can be set by hand. Next to them, down the line, is an array of mechanical switches, perhaps six feet high by ten feet or more long. Along the way are a couple of paper tape readers, and then the monster trails off into a few more gadgets, and all of this apparatus is interconnected into one fantastic Rube Goldberg machine. The whole device is twenty feet long, at least, and it represents no more than one-third of the original beast: Commander Howard Aiken's Mark I, an electromechanical device that some claim was the first true computer. Aiken and his co-workers designed it before the Second World War, and IBM, then a manufacturer of punch cards and card readers, built it to his specifications and delivered it to Harvard in 1943. First used to figure out ballistics for cannon fire, by the end of the war it performed some of the calculations that a group of researchers at Los Alamos needed to design the firing mechanism for the atomic bomb. The Mark I was a sluggish device—it could take more than two weeks to finish a complicated calculation—but it worked, and even at its snail's pace was far faster than any human being pulling the lever on an adding machine.

To anyone of an age that the world has always been filled with computers, a world, in fact, in which computers have become household toys, the Mark I in all its glory seems unbelievably cumbersome, as much a part of history as the Pony Express or Eli Whitney's cotton gin. Although the computer has been in use for barely a generation, it seems far older primarily because it has transformed science almost beyond recognition. Well before the first computers were ever built, the scientific need for them was already clear, particularly in climate science; without computers the weather would never yield to scientific inquiry.

In fact, long before the Mark I ever calculated the trajectory of an artillery shell, the First World War had provided the same incentive as the Second for scientists to come up with better forecasting techniques. Just after the Great War, the British physicist and psychologist Lewis F. Richardson tried to develop a method of calculating at reasonable speed certain equations that could describe the motion of the atmosphere—thus allowing him to predict from a set of initial information what the weather would do next. The calculations involved, however, were extraordinarily complex; what makes the task of fitting the atmosphere into the basic laws of physics so difficult is that so many things are going on at once with the changes in pressure, energy transfer, wind, precipitation, and so forth; all of which have to be accounted for in some fashion or other if the sums are to come out right. Lacking even rudimentary programmable computing devices, Richardson proposed a fantastic kind of forecasting factory:

Imagine a large hall like a theatre, except that the circles and galleries go right round through the space usually occupied by the stage. The walls of this chamber are painted to form a map of the globe. The ceiling represents the north polar regions, England is in the gallery, the tropics in the upper circle, Australia in the dress circle and the Antarctic in the pit. A myriad of computers are at work upon the weather of the part of the map where each sits, but each computer attends only to one equation or part of an equation. . . . From the floor of the pit a tall pillar rises to half the height of the hall. It carries a large pulpit on its top. In this sits the man in charge of the whole theatre; he is surrounded by several assistants and messengers. One of his duties is to maintain a uniform speed of progress in all parts of the globe. In this respect he is like the conductor of an orchestra in which the instruments are slide rules and calculating machines. But instead of waving a baton he turns a beam of rosy light upon any region that is running ahead of the rest and a beam of blue light on those who are behindhand.

The "myriad of computers" Richardson sought were people, 64,000 of them, he estimated, working out the equations by hand, cranking out their numbers on office adding machines and then passing on their results to the next person in line.

Richardson, sixty years ago, had the right idea. There are five equations—mathematical expressions of laws of physics—that taken together can describe motion in the atmosphere: how strong the winds are, where they blow, how they change. One is Newton's famous Second Law of Motion: $F = MA$, force equals mass times acceleration. This basic equation can reveal just how the winds respond to any of the stimuli that could set the atmosphere in motion, such as the switch from day to night, the rotation of the earth, changes in pressure, and so on. Another is the First Law of Thermodynamics, which accounts for the effects of heating and cooling in the atmosphere. A third equation describing how mass is conserved in the atmosphere allows us to depict mathematically the fact that if winds blow into a part of the atmosphere, air is going to have to move out of that area, or the atmospheric pressure there will change. The fourth, called "the equation of state of an ideal gas," relates the temperature of the air to the atmospheric pressure. The fifth equation estimates how fast the vertical winds are blowing.

That's all that's needed to produce a mathematical model of an atmosphere—five equations, none of them with more than three terms (or four at the most, depending on how they are written). All of them involve simple, basic concepts, some of which are centuries old and all of which are at least mentioned in high school physics classes and taught in some detail in a first college physics survey course. Beginning with a set of numbers that describes the current temperatures and pressures at selected points in the atmosphere, we can, in theory, plug the numbers into those five equations and solve them all at once to see what the atmosphere will do next. Repeat the process, taking the results from the first go-round as the starting point for the second, and so on, and the calculations can carry forward to forecast the weather for a day, or two, or three.

There is a catch, however. The fundamental equations are simple enough, but in fact each of their variables hides a ream of other calculations, all of which have to be figured out before reaching a value to plug into the simple equation. In the thermodynamic equation, for example, you must account for the adding or subtracting of heat over time. If you want to account for the effects of a greenhouse gas, like carbon dioxide, you have to figure out a set of equations that describes what happens when you add or subtract carbon dioxide from the atmosphere. If you want to explain the effects of volcanic dust, you have to quantify both the dust's scattering of sunlight and its absorption of solar energy. And each of these equations—and all the rest needed to add greater detail and accuracy to your mathematical approximation of the atmosphere—has to be calculated on each run through the problem. What's more, they all have to be solved repeatedly for each relevant location on the earth. The more points on the map you use, the more accurate your picture of the atmosphere will be, but accuracy is costly: the number of calculations you have to do shoots up once again. Today the simplest weather models take about one billion mathematical operations to produce a single day's forecast. Richardson, even with his extremely simple model, hadn't a prayer.

The great advance that distinguished the first true computers, built during and just after the Second World War, from calculating machines available even in Richardson's time was more than just the improvement in speed. These new machines could be programmed; they could store both instructions and the results that were generated from a calculation. Within the first few years after the war, a few scientists outside the ranks of those who had harnessed the computer to wartime tasks seized upon the new machine, and the first major problem they tackled was modeling the atmosphere. During the next five years, a very small group of researchers managed to produce a weather report by calculation instead of by matching map to map. Once they demonstrated that a mathematical model could equal or beat the accuracy of the best forecasters, they had built the structure upon which were hung all the later bells and whistles: adding oceans to atmosphere models; considering decades of climate instead of days of weather, and so on.

Philip Thompson played something of Dean Acheson's role throughout this process—he was present at the creation, and had a hand in moving it forward as well. He is today a tall, quiet man, uncomfortable talking to strangers given to abstraction. He holds himself straight, almost at attention, and he looks, as he is, the very model of a retired Air Force colonel.

As he tells it, though, it was rather different in the 1940s, when he was a lieutenant on active duty stationed at the University of California at Los Angeles, assigned to analyze weather maps in great detail, calculating pressure changes and the like. Thompson's calculations, unfortunately, produced useless results: he generated errors as large as 100 percent. So he abandoned the official objectives of his assignment and started working, first on the back of an envelope, then with an old Monroe hand-crank calculator, to produce a set of equations that would correlate wind speeds with atmospheric pressure. By early in 1946 he had a fairly complete set of equations, but it was, Thompson says, "a good thing that I knew nothing of Richardson's work": Richardson had figured out the same equations at least twenty-eight years earlier.

In the spring of that year, though, Thompson came across an article in the *New York Times Magazine* that told of an effort at Princeton, lead by John von Neumann, to build a calculating machine to be applied specifically to predicting the weather and measuring the consequences of human intervention in the climate system. So Thompson trekked east by "B-29, bus, stagecoach, train, oxcart, and the PJ&B [Princeton Junction and back]," met von Neumann, impressed him, and managed to get reassigned to von Neumann's lab quickly enough to leave him just time to retrieve his clothes from California. Even within the context of von Neumann's group, the magnitude of the labor was still daunting. "I quickly realized that this job was more than one small boy could handle," Thompson says, so he began staying up late at night, drinking a lot of beer, picking away "at the periphery of the problem, trying to understand the different kinds of motion that occur in the atmosphere." Then, in November 1947 Thompson received a letter from an old friend, Jule Charney, who told Thompson that he had figured out how to represent mathematically certain kinds of winds in a way that a computer could be programmed to calculate, and then asked for a job with von Neumann.

Charney got his job, and as a result of his mathematical breakthrough, the machine builders came to lag behind the meteorologists. Von Neumann's computer, dubbed the Johniac, wasn't ready when his colleagues were set to program their equations into a machine, so they used the ENIAC device, then the only true electronic computer, programming by moving plugboards and operating by hand an enormous bank of keys. Finally, in April 1950, they ran on the computer a model of the earth's atmosphere in which winds blew, pressures rose and fell, and the weather changed, completing a day of weather in a few hours within the machine. Producing a picture today of what the weather would be like tomorrow, the machine could, albeit imperfectly, predict the future.

A black-and-white photograph was taken of the celebration that followed. It is a conventional shot, eight men in suits and ties, long jackets and baggy trousers, all standing in a row. A few of them are making a pretense of conversation; Jule Charney stands at one end, and on his face more clearly than on those of his colleagues is the look of the cat that ate the canary. Except for Charney's expression, the picture has the air of a hundred thousand other such gatherings, where people stand still for the camera at the end of one meeting or another.

But there is that smile on Charney's face. What he and von Neumann, Thompson, and the rest had done was to create a new kind of machine, a novel, hybrid mechanism that in its turn created an entirely new place in which a scientist could work. During the war, the computer had been used as a fantastically fast adding machine. The combination of the computer and the mathematical model produced something more: a flexible machine, one that you could modify as you refined the terms of one equation or another, one that could create as many different simulations of the atmosphere as you had the wit to build. The researchers could add more sunlight than the world receives today, or take some away. They could, in time, build planets made all of land, or all of ocean; create a world with no ice at all; cover the globe with ice; add carbon dioxide; take it away—whatever they chose.

Here was the key to the revolution. Before the Princeton group fashioned its models, there was no experimental science of climate. Previously, researchers could describe what they saw happening—take the experiments nature dropped in their laps, as it were—and uncover some of the mechanisms that drive the interactions between land and sea and air and life. But they could only watch, measure, and think about what they saw. With the new models they could play. They could see what happened as they set their planets in motion. In their sober moments, they could follow the classical strictures of the scientific method by changing just one variable at a time and running an experiment to gauge the role of each cog in the system.

What Charney and his colleagues did was to liberate climate science from the intimate, confining embrace of nature, placing an intermediary between the scientist and the intractable complexity of the real world. Of course, the model worlds within the computer must be checked against reality to ensure that the conditions that actually result in rain, for example, also produce precipitation within the machine's simulation of current atmospheric dynamics. Apart from that bound, though, the scientist was free to roam through any number of different simulations, different imagined planets, within the computer, and finally back out again with conclusions drawn from the model experiments that can dissect the history and daily behavior of the climate we experience around us now.

The new science of climate—these multiplying novel worlds—needed above all more computers, bigger ones, swifter ones. And such machines needed a home. By the late 1950s, the powers that were—the National Science Foundation, the universities, some large corporations—came to the conclusion that the computers used in atmospheric studies ought to have one home, a national laboratory that would become the core of American weather and climate research. The original name for this establishment was to have been the National Institute for Atmospheric Research, until someone pointed out its acronym spelled backward was "rain." So instead the proposed base for climate science was called the National Center for Atmospheric Research (NCAR), and Walter Orr Roberts was asked to build it.

The corner of the world that most often plays home to Roberts is Boulder, Colorado. The city of Boulder runs slap up to the edge of the Rockies and stops, by design. Drawn across the city's maps is a blue line that traces the contours of the rise at the western edge of town, just before the mountains rise straight up. The city won't pump water up above that line, and in the West the threat of waterlessness is enough to stop even urban sprawl in its tracks.

The mountains, thus, dominate the town. The Front Range of the Rockies, a fairly new formation, rose out of the sea that covered the Midwest during the Cretaceous period some 70 million years ago. The core of the Rockies is much older, however, constructed of granites that formed at least 1.75 billion years ago. As the newer range lifted up, the old core rock of the mountains flexed and cracked. Consequently, a small fault line runs along the edge of town. There a series of small peaks was created when slabs of rock lying on their sides were tipped upward, to angles of more than 50°. They look like what they're called, the Flatirons. Beneath them lies a broad and gentle hill, topped by the flat crown called Table Mesa. A herd of deer often grazes the mesa, and at least one red-tailed hawk calls there frequently during the winter.

A single major building stands on Table Mesa. It was built of a dull pinkish concrete, and though its two towers stand five stories tall, in the right light—the light of afternoons, in the summertime—the building seems to vanish altogether, lost in the background of mesa, pine, and mountain. This building, the home of NCAR, is situated well above the blue line; it owes its existence, in large part, to Walter Orr Roberts's having persuaded the citizens of Boulder that it ought to be built.

Under ordinary circumstances, the building of a laboratory is of no great import, just part of the normal business of science. After all, science isn't an abstract blend of theory and experiment, but rather theories and experiments worked out by people sitting in offices or moving about along a lab bench, in facilities requiring plumbing and light and heat and all the rest. But in times of revolution such ordinary facilities take on extraordinary meaning. A new synthesis of theory, a new mode of experiment—these climate science had by the early 1950s. What it lacked, the third component of a successful intellectual breakthrough, were the institutions needed to foster the new studies. Just as the disciplines from which climate science sprang had and have their own concerns, so the institutions in which the disciplines are studied and taught—university and research labs, think tanks—have little time or, more important, little money to hand over to people who persist in studying questions off the main track of physics or chemistry or what have you. To assert its independence, back at the dawn of the field, climate science required a room of its own.

There are two ways to create such institutions. One is that undertaken by Isaac Newton, who waited until his elders died or slowed down, and then from his position as head of an existing organization, the Royal Academy, fostered what he saw as essential to British science. The other is to create anew, to build not just a physical structure but a society of people committed to some version of the same broad enterprise, a group that can recognize success on its own terms and can affirm that the issues with which it is concerned are of central importance. It was an act of revolution to create NCAR, and Roberts was handed the task of transforming a revolution in thought into concrete and glass on a hill above Boulder. In this he succeeded, beyond question.

On the wall in the main entranceway to the NCAR headquarters hangs a plaque. It carries a picture of Roberts, taken some years ago, when he had more hair than now, and a list of some of his positions and honors. That catalogue of his achievements ends with a Latin motto, which, translated into English, reads "If you seek his monuments look around you." A phrase of convention, virtually a cliché, in this setting it retains the force of its original meaning. NCAR: the site, building, and institution contain reflections of Roberts's personality, of what the man thinks is important in science.

For Roberts, the setting itself establishes the proper context for thought. When he undertook the job of creating NCAR, Roberts owned a home with a view south, toward Table Mesa; he lives there still, and when he sits in his living room, he can see the structure he created. For years before there was any prospect of a climate laboratory, he had watched the play of light across the mesa, walked on the hill, and had in fact tried once to establish a technical institute there. By the time the construction of NCAR faced him, he knew precisely what he wanted. Roberts persuaded the state to give the mesa to the project. And then Roberts turned around and, in effect, gave the site away as a park for Boulder. He persuaded the governor to grant the land only on condition that NCAR bury its power lines. Then he convinced the NCAR trustees never to close the grounds except in times of war so that the land would be left undeveloped and open. There is today one exception to the blue line, agreed to by referendum in the town; in 1961, the Boulder voters, persuaded, at least in part, by Roberts's fierce protectiveness of the mesa, granted water to NCAR.

And then he built his laboratory. It looks—and the resemblance isn't an accident—like a castle perched upon the rock. Its two towers stand at the northern and southern ends of the complex, with smaller promontories standing on top of the main keeps. Its design echoes the Moorish palace of Alhambra, and it has as well a recognizable kinship to the cliff dwellings at Mesa Verde. When the committee guiding Roberts hired an architect, they chose the young, relatively unknown I. M. Pei, and the building he produced looks like no other laboratory: it is beautiful, a sculpture atop the mesa.

The beauty was not achieved, however, without a struggle. "I rejected his first design," Roberts says. "He was a little bit upset about that. I said, 'It's just too tight, too concise, too compact,' and we sent him back to the drawing board to spread it out." Roberts wanted it to be impossible to tell how many stories high the building was, so there are no horizontal lines breaking up the towers. And he ordered booby traps worked into his own building:

I wanted a building that looked as if it had been designed for something else and got remodelled. I wanted an individualistic character for each of the offices. To me diversity is a great strength in a building, just as in ecological systems. I made it a requirement that if you were going to go from point A to point B in the building, there had to be at least four routes and no preferred one. If someone at the front door wanted to go to someone's office, I wanted them to need a guide.

Roberts's own office is relatively easy to find: walk through the main entrance, turn right, then left, go up one flight of stairs, turn right, go almost to the end of the hall, and then just hook around to your right once more. His window looks over one of his and Pei's mistakes—a fountain now permanently dry. The two men had forgotten that quite often during the winter the winds gust up over 100 miles an hour, gales strong enough to turn a fountain into a driving rainstorm.

Roberts did not make many such errors, and he was on guard against one trap in particular. The great danger in any revolution is that the revolutionaries, once successful, will revert to the behavior of the overthrown regime, as, for example, the KGB appropriates the legacy of the Tsarist secret police, or the Chinese Communist Party reproduces a model of bureaucratic life to rival any contrived by the Manchu emperors. In climate science the analogous reversion would have been to re-create within the field the separate disciplines from which the initial ideas flowed, to wall off, for example, the physicists from the chemists. It is always a hard-to-combat tendency. Roberts was aware of the risk, and built accordingly.

Henry Adams explored in *Mont-Saint-Michel and Chartres* the proposition that architecture is a mirror of thought and feeling, that a structure not only reveals but molds the theology, philosophy, and emotional reach of its builders. In Saint Thomas Aquinas's effort to reconcile God's omnipotence with human free will, Adams sees a mirror of the architecture of the day, Thomas's theology becoming, for Adams, an exercise in high Gothic art. The

NCAR building has a similar impact, revealing at least Walt Roberts's idea of what the practice of his science ought to entail: "I'm quoting someone else," he told me, "but what I believe is that the atmosphere begins at the sun, reaches out through space, includes the very top layers of gas around the earth, reaches down through the atmosphere to the land, and then from the surface of the oceans down to the ocean bottom." In a word—everything.

So as a practical matter, the design of the building enforces the idea of collegiality. People on the mesa must encounter one another, talk with one another; an atmospheric physicist cannot avoid bumping into the occasional chemist or storm-system modeler; someone who flies balloons to collect information from the stratosphere will at least from time to time rub shoulders with someone whose central concern is with the dynamics of ice. And as an assertion of belief, the NCAR structure, with its convoluted paths from here to there, its isolated site, and its air of being a city on a hill, is the concrete expression of the abstract, ultimate ends of a new science, one that recognizes no divisions from the surface of the sun to the bottom of the deep blue sea.

Roberts himself is no longer an active researcher; when he took the job at NCAR, he had to leave the lab floor, and too many people had accomplished too much for him to catch up after he left his post in 1973, after twelve years on the job. And NCAR is by no means the sole locus of climate research in the United States. NASA has several groups doing climate work, the Princeton group created an extremely successful institution called the Geophysical Fluid Dynamics Laboratory, and a host of other centers have been opened at universities and other labs. As far as bricks and mortar, funds and tenured jobs, and the other material underpinnings of the research effort go, the revolution is mature, consolidated within the community of science. But NCAR stands as Roberts's monument. The sign on the door states it, his colleagues honor him for it, and the place remains, in its structure and the way in which the institution compels a view of nature that crosses more conventional boundaries, a mirror and an image of what the new science strives to do.

Climate, the actual climate out there in nature, has been transformed by the science that studies it. It has become since 1945 an integrated system, a weave of processes whose workings spread through time, over billions of years, to produce the world we live in. The science expanded the concept of climate to make it reach out through nature, from the sun to the bottom of the sea and back again. The concept of climate has also transformed the science and the scientist. It has replaced weather historians with a tribe of Manicheans; makers of worlds; scientists who must slip from ocean to ice to atmosphere; scientists who must, by the nature of what they aim to study, integrate knowledge from all over the map.

Ultimately it has forced a revolution in the way these scientists must do their science. They cannot simply look at nature, make a measurement, perform a test, and announce their result. They must take nature, abstract and simplify some model in their minds that matches what nature looks like, build that model in a machine, and then, finally, begin to see if their conception has any connection with the real world that they wish to understand. As much as the concept of climate has changed, so have the means with which people arrive at that concept. As the climate itself comes to seem more interconnected, the models become more complex, and each refinement in one forces a response in the other to the point where understanding why a storm occurs in a model atmosphere can be as difficult to figure out, Roberts says, as what causes a gale to blow through the real Pacific some fierce afternoon.

And this change has happened in the course of a lifetime. No one man made this revolution, no giant like a Newton or an Einstein. But occasionally it is possible to recognize the central characteristics of the enterprise by following the career of a single figure whose life in the science spanned the crucial events that forged a new line of inquiry. Roberts, years ago, in a Graham Paige broken down in some town with no name; Roberts today in his armchair in the living room that looks toward Table Mesa; or Roberts leaning back at his desk in the building that is his monument with its fountain that must remain dry. There are layers of chance that intrude on the journey from there to here: the happenstance of a bad automobile cooling system; the fact of the Second World War; the serendipity of his jilting by one girl (and staying on the East Coast) and waiting to wed another (and trekking west). Then there is a counterweight that rigs the wheel: Roberts himself, his sense of where the important questions lay and lie—the knack.

CHAPTER 8

The Machine's Eye

A T THE BACK of the lobby at NCAR a large window is cut into the floor overlooking the computer center for the laboratory. Tourists, hanging over the railing that surrounds it, look down on the heads of the computer operators and try to get a count of all the different anonymously sophisticated-looking metal boxes throughout the room. Down in the center, directly below the window, a bank of display screens is monitored on the fly by one or two people. Each of the screens monitors the status of one of the major systems in the room, telling the operator what jobs are running, how long they are taking, whether any problems are beginning to show up. Atop one of the monitors is a small digital display that continuously flashes numbers—5.6, 8.3, 2.4, 10.0—that indicate the wind speed on the mesa just above and outside the computer center. During wind storms, those speeds can reach 100 miles an hour, and occasionally accidents happen: the power sometimes fails, and the machines in the operators' care may crash. When that occurs the staff must obey a strict rule: Watch the wind gauge and wait; turn nothing back on until the winds drop below 50 miles an hour. From below, the window in the floor becomes a skylight. Lightning is common enough on the mesa, and a strike close enough to home can fry millions of calculations, not to mention transistors. If the power fails during a thunderstorm, the staff on the floor of the computer room can look up, away from the machines that contain the electronic winds of a model's weather system, and watch the lightning from a real storm play over the mesa. What goes on inside the machines are imitations only, schematics of storms, line drawings of the weather. Outside, overhead, the real thing dominates. The occasional failures of the computers become a kind of reminder, a relatively gentle warning: for all the power of the latest computers the weather itself remains in some way intractable, beyond control. The computer operators are used to the contrast. After a strike they simply wait patiently until the thunderheads pass by. Then and only then do they power their machines back up.

Spring 1986: Just in front of the bank of monitors and a little over to the left stand two machines that look faintly like the circular sofas you can still find in Victorian hotels, the overstuffed, high-backed kind that used to go around pillars in a lobby. These devices are doughnut shaped but have one wedge of the circle cut away. The outer ring of each, which is just high enough to sit on, is covered in black vinyl, while the back of the sofa, the inner ring, rises in flat panels to a height of about six feet. One of these inner sections has had its metal panels replaced with glass. From inside the hole in the doughnut nothing is visible but a tangle of blue and white wires; from outside the circumference of the machine, you see slab after slab of copper. The machines, each of which is perhaps five feet in diameter, are far from the largest devices in the room and very far from the noisiest; in fact, with no moving parts, they are completely silent. They are the reason that all the rest are there. The sign above them names them: Cray-1As. They crunch numbers, unbelievably quickly, in huge, voracious quantities.

The creation of atmospheric models for weather prediction that began in the 1950s culminated by the 1980s in the development of climate models, whose central purpose is to provide an electronic laboratory. Climate theory is embodied in the equation chosen for a particular model; calculations in the computer provide the experimental test of those theories. The enormous number of calculations that these sophisticated models rely on requires the largest available computers, or so-called supercomputers. The Crays are supercomputers, a bit long in the tooth now, but still able to handle the large models that NCAR has made its stock-in-trade.

The ultimate goal of the science of climate is to understand the planet's behavior sufficiently to predict its evolution and to gauge the effects of human action on the habitability of the earth. What distinguishes climate science from older traditions of research is this dependence on the computer. A model is an abstraction; the degree of verisimilitude achieved in any simulation depends in part on the skill of the modeler and in part on the capacity of the computer. In every experiment in climate theory, the computer and the model interpose themselves between hypotheses about nature and the accomplished facts of nature that, finished and bound, storm across the top of Table Mesa on a winter's afternoon.

In other words, the computer-based scientist must leap over a second hurdle, one more than his more traditional counterpart. Model-based science forces its practitioners, before they can understand what they see in their results, to dissect how they constructed the thought, the electronic experiment, which allowed them to produce results. The study of climate becomes, in large measure, the study of climate *models*.

Within the field there is a hierarchy of models, beginning with very simple, obviously unrealistic schemes up through simulations that grow increasingly complex. A one-dimensional model, for example, assumes that the whole world is a single point, made up either of land or sea. It averages global temperatures into a single value for that single point and then examines changes only up and down a vertical line, from sunlight entering the top of the atmosphere down to infrared radiation leaking up from the bottom.

The next step leads to two-dimensional models, which increase the amount of detail in the climate by adding climate events and winds that vary along one axis. Where one-dimensional models produce a line extending vertically, two-dimensional ones sweep out a vertical plane of the earth along one line of latitude or longitude, a world on a disk. Three-dimensional models mimic globes: they integrate motion over the entire surface of the earth. These so-called general circulation models include simulations of the great flows of wind that move air, heat, water, and trace elements along for the ride throughout the atmosphere.

Three-dimensional models are presumed to be the most realistic; they are certainly the most complex. And they are all very much alike, in broad outline at least, which is no accident. The NCAR version, called the community climate model, is the standard for American climate research, and it is available to any scientist in the climate-research community. Other labs have borrowed and modified it, and in its current incarnation it is itself the rewrite of a

British modification of an Australian original. All models are cumbersome to use; no longer is any one of them written by a single scientist.

The very complexity of these higher-level simulations often makes their simpler counterparts attractive. Some extremely sophisticated problems, for example, the issue of the role of plant respiration in a hydrological cycle, lend themselves to one-dimensional models. A first attempt at such an issue would involve looking exclusively at a vertical column of water as it travels into the atmosphere from rain, through soil, and up again through the leaves of a plant. But to study in detail how the climate system behaves, one must use the most comprehensive simulations available. Testing ideas about the transition in or out of an ice age, for example, requires a laboratory, a model, in which events can run out over the time scales necessary and in which the climate regime being studied can evolve across the globe, so only a three-dimensional simulation will do. At the same time, the availability of models of particular sophistication shapes the choice of a problem to study. It is impossible to perform an experiment about the global carbon cycle without possessing or creating an environment in which the various carbon sources and sinks can interact, and the "environment" here is the electronic one of a very complex system of equations. Climate science and the limits of climate research at any time are defined by the limits in the sophistication of the models available.

In the abstract, those limits are set by the abilities and creativity of the scientists composing their systems of equations. In practice, the models are circumscribed by the capacities of the machines available. A one-dimensional model can run quite happily on an IBM personal computer. A three-dimensional model consumes enormous quantities of processing power because the simulation of a single year's weather can require several hundreds of billions of calculations or more, depending on the model's precise details. Without a machine swift enough to handle it, a question of climate science that hinges on global interconnections remains unanswerable, no matter how elegant the original conception. The workings of a machine begin to delineate the parameters of the modern science of climate—what drives it, how it chooses its territory, its areas of concern, what answers it may hope to get now, how to judge the quality of those answers, and what the model may soon achieve.

Historically, NCAR has taken full advantage of the machine's potential in climate science, pioneering the use of the extremely large computers that began to come on-line in the late seventies. NCAR's supercomputing capabilities began with the operations run by Gary Jensen. Jensen is a man out of the common run at NCAR: Within the center Ph.D.s sprout like dandelions; Jensen never attended college. NCAR staff members have caught the fitness bug with a vengeance: the twisting road up the mesa is made more hazardous by a stream of animate obstacles—runners struggling up the hill, suicidal bicyclists swooping down. There is what New Yorkers call a "granola bar feel" to NCAR; many of the people there look to be the sort who eat whole-grain breads and drink unstrained fruit juices. Jensen has something of the appearance of the men you see sitting way back on motorcycles that run on automobile-sized engines.

Jensen is nonetheless enormously skilled as a computer hardware man. Before he came to Boulder he built the center that NASA's Jet Propulsion Laboratory used to process the information from the Voyager missions to Jupiter and Saturn. At NCAR, Jensen takes an enormous and obvious pleasure in being the man in charge. Walking over from his office to one of the Crays and pointing to the serial number 3, he says, "We're in all the textbooks because of this machine. It's the first production Cray. Serial number 1 was the prototype—it was half as fast, half the size. Then Seymour Cray built number 2—and it didn't work. So they built number 3 with parts from 2 and they sold it to us." What they wound up selling was the box, and nothing more. The Cray corporation had to buy back half the time on the machine for six months just to write the programs that allowed it to determine what its computer could do, which meant that NCAR had to figure out from scratch how to make the machine work for its own purposes.

What made that task particularly difficult is that the Cray is essentially a terribly uncomplicated device. It simply adds, subtracts, multiplies, and divides. It is a racing machine, designed purely for speed. Most computers have some form of switch that acts as a timing device to control the flow of information from one of its parts to the next to ensure that the information being transferred doesn't step on something already going on in the target location. The Cray lacks such switches. Instead, each wire leading from one location to another on all of the boards that make up the computer is precisely measured, and movement in and out of locations in the Cray's memory is regulated by the length of wire each signal must travel, so no time is wasted waiting for a switch to open or close. There are 60 miles of wire inside a Cray (down from 6,000 miles in the CDC 7600, the former state-of-the-art computer). The idea behind the Cray was to get those wires as short as possible (hence its circular shape) so that in effect the only limit placed on its speed was that imposed by the speed of light.

The raw numbers that result are impressive, though almost impossible to grasp. A Cray-1A can execute 80 million instructions every second; properly programmed it is capable of around 200 million instructions a second. The longest it takes the Cray to transfer a piece of information from one location within its memory to another is one-billionth of a second. A rough estimate as to the time it takes a Cray to perform the average job at NCAR is ten seconds of central processor time. In those ten seconds the machine will scoop up 500 million pieces of data to process and will spit out another 500 million pieces as the result. It takes eighty white pines each day to supply the paper needed for the printouts that the Crays spew out.

The meaning of such astonishing numbers? Some comparisons: the ENIAC machine, used to make the first numerical weather forecast, could manage 5,000 calculations a second. A model that runs in a minute on the Cray will take forty-five minutes on an IBM mainframe computer, an hour and a half on a middle-sized DEC minicomputer, and two and a half months on a stock IBM PC. NCAR's Crays manufacture numbers faster than any of the other machines the center possesses, and they are making their numbers second by second, twenty-four hours a day, year-round, barring breakdowns and the occasional routine tune-up.

Once you have your 500 million words of data to process, or your half billion results, you have to do something with the information. The machines that surround the Cray exist to channel, organize, and finally use this unceasing river of information. A supercomputer remains dumb without them, its gatekeepers, runners, archivists, scribes, even guards. The complexity of the models that have been developed over the last decade has created the need for supercomputers, the Crays and their successors that are now coming on-line. The capabilities of those machines have in turn fostered a novel style of computing, one in which problems are solved across a system, an interconnected web of devices. A climate model is a metaphor, rigorously defined, of the actual features of nature, the climate experienced outside of NCAR. The computing system that has emerged over recent years to manipulate such models has itself evolved to resemble those models in certain features. In particular a supercomputing system possesses, as an analogue to the atmospheric flows that the models mimic, its own currents of information, which merge and divide, interact, and, at times, clash.

Tnus a room full of metal boxes becomes, at least schematically, an entire planet, one through whose peculiar geography the scientist must learn to navigate. The scientist Gloria Williamson developed one model to simulate a full year of the globe's climate. When she wants to perform a single run of her simulation, she must first gain access to a relatively small computer before entering into the NCAR network that ultimately leads to the Crays. A number of computers scattered around the building can do the job, and after selecting one Williamson tells it which model she intends to use and which set of weather data she wants as the initial conditions of the run. Often, depending on the intensity of traffic to and from the Crays, the computer she enters then calls on another, larger one (an IBM 4381 in this case), which acts as the main gatekeeper for the Crays. The IBM slots Williamson's job into a queue and then waits. Almost no one goes straight on the Cray, except the systems programmers. When the job finally reaches its turn on the supercomputer, the IBM also passes on the list of data Williamson has specified that the Cray will need from the library in order to perform the task. As soon as the Cray gets the job, it calls for that information and then immediately unloads the job right back out onto one of its disks. The central processing unit is never allowed to stand idle or to wait for data from somewhere else, so different jobs are continuously rolling in and out of the machine, with each model having to go back and wait in line each time it requires more data.

Over the past twenty years, however, NCAR has accumulated a

library of 14 trillion pieces of data, with the size of the data store doubling every year. Some person has to walk over to a bank of cassettes, locate the right one, grab it off the wall, and load it by hand into a machine that can read the tape and transmit it back to the supercomputer, which is capable of 200 million operations a second once it has something upon which to operate. That is the reason jobs are rotated in and out of the Crays, inefficient as that might seem; any other course would leave the Crays idle while the slower components of the supercomputing system, including the human component, struggled to perform their tasks. In addition to all the initial information needed to start up a model once it reaches the Cray, a model continuously generates and then requires data as it runs; the results it produces from interim calculations must be stored and, when the next iteration of the model demands them, recalled. For Williamson's particular model—a relatively information-hungry variant of NCAR's main large-scale climate simulation—the Cray routinely must wait for forty-five minutes to get its data; if it is a busy day, it may take the operator two hours to load and adjust all those tapes.

So after aeons, subjectively speaking, of waiting on the right data, the Cray receives the necessary data from the computer that runs one of the mass storage systems. When the Cray has all the information it needs, it retrieves the model and loads its data into the central processor, performs a calculation, spits out the results, and then rolls the model back onto the disk while it transmits the results to a history tape that stores intermediate calculations. While the mass-storage device's computer processes the creating and updating of the history tape, the Cray resumes work on another job and performs a number of instructions until the rest of the system catches up with it. Then it hauls the original model back in to perform the next calculation. Williamson's model requires three hours and forty minutes of the Cray's central processor time; factoring in all the time required to ferry information around the system, it actually takes about thirty-six hours to complete the calculation.

Only then, after a day and a half, may Williamson actually see what she has wrought. If she chooses, she may make a movie of her year. The statistics that are the first product of a model run can be used to generate a variety of different charts and graphs, including the familiar outlines of weather maps, flat projections, or circular representations of a hemisphere. Another machine in NCAR's arsenal will take each of those maps for every interval that the modeler specifies and trace them on 35-millimeter film to create, in effect, a moving picture of what the weather might be. Most of the data, though, is never seen in any form; a few percent, at most, of the results from the calculations on each step of the way get written onto a history tape. The rest are created and then used as the model works through all of the intermediate steps needed to produce a single temperature reading at a single point, for example. The only time that information is preserved and examined is when the model appears to have within it a bug, which may lie buried within some operation a dozen steps away from the model's output. Then the would-be exterminator has no option but to seek it out through all the interminable numbers that the Cray generates on the way to producing its ultimate simulations.

The way many bugs show themselves first is by blowing up the model, by requiring the Cray to perform some operation that overtaxes either it or the devices that support it. Whenever problems do occur, Jensen is the court of last appeal. Programs may contain an endless loop, sending the Cray to do the same operation over and over again, or the poor construction of the way they channel their data may threaten to fill up all the disks that NCAR possesses. Whatever the particular bug, it will ultimately cause the machines to crash, causing all the jobs then running on the supercomputer to be lost. The Crays issue a warning, Jensen says, "just before they're about to run up the white flag and give up." When that happens the operators, on standing instructions, dump the erring job, destroying whatever it was the offending program was trying to do—all this with a conspicuous lack of sympathy for the scientist whose work gets trashed. (Welcome to the major leagues, kid.)

Such disasters serve to train those who ultimately become the adepts of what is, after all, an unusual method of doing science. To get worthwhile results out of models requires not simply a knowledge of how to build a reliable, accurate, mathematical picture of natural processes but an understanding on a very intimate level of how to construct that system of equations so that they may move with the least fuss and bother through a network of machines that each has its own peculiar characteristics. Crashes are the spectacular consequences of failures to understand the silicon substrate of climate science; it is more common to find people who are simply inefficient, whose work proceeds slowly, who get less done, or must ask less complex questions because they are unable to divine the path between model and machine that will move their experiment most swiftly to a result. Sciences based on observation or direct experiment do not face this. Crudely, if you want to test the effects of cyclamates on rats, you feed cyclamates to rats, wait a while, kill them, and see whether or not any of the cadavers was developing cancers, or whatever it is you may be looking for in whatever system it is you wish to examine. Sciences based on models, and none is more model-dependent than climate science, require their students to develop new habits of thought, habits grounded in the actual architecture of the machines they use to create the model environment.

This is what makes the statement that climate science is machine driven something other than a commonplace. Lewis F. Richardson failed in large measure because he lacked the technology to execute an idea. He, or his network of human "computers," could not add, subtract, multiply, and divide fast enough to produce a prediction of the next day's weather before the next day dawned. But the webs of machines now in use are far more than simply superfast adding machines. The very simplicity of them, the fact that the Crays actually do merely add, subtract, multiply, and divide, has produced an extraordinarily flexible laboratory within which climate theory is now developed. When properly programmed, that is, directed by a well-conceived model, the modern generation of machines can be driven to simulate almost any climate pattern that a scientist can imagine.

The product of the union of the model and the machine is a representation in silicon, copper, and electrical signals of an entire planet. Usually the scientist can only see the computer's geography after it has been transformed into the familiar geography of the seven continents, but the machine retains its own landscape, its own courses from here to there. Bugs and crashes are the penalty for poor navigation. Climate science advances not only when faster machines come along but also when researchers delve more deeply into the shape of the model worlds they create, a shape dictated by the structure of the machine within which their simulations exist.

NCAR plays host to several of the best such explorers-in-silicon within the American climate science community, one of whom at least has found in NCAR's peculiar style enormous freedom to move through the multiple worlds of nature and of models. NCAR publishes a pamphlet that describes its building and the reasons that underlie different choices in its design. In the corner of one page is a small photograph that shows a heavyset, bearded man sitting, staring at a computer printout that trails off his lap and onto the floor. The famous Steinberg cartoon of the world as seen from New York hangs on the wall; every other surface is covered with a mountain range of papers. The caption reads, "Pei designed the building's varied office spaces to give scientists the freedom to be messy." Messy? Bob Chervin's office goes beyond messy into legendary; it is an active demonstration of the concept of entropy.

Chervin, among his peers, is something of a lone wolf. "I use an analogy," he says. "In mountaineering you've got two basic approaches. You can mount an expedition with 200 porters and 50 Sherpas and try to get to the top. Then there is the alpine approach: You and your buddy put packs on your backs and start climbing." The people around him have the job of building the NCAR's standard central model. To Chervin that standard model is what a stock Harley is to a serious biker: raw material to be molded into shape.

Once, as Chervin tells it, he was pure. As a graduate student he studied plasma physics; his thesis was a highly theoretical piece of work that never once gave him cause to touch a computer. Now, as he tells it, "I am totally contaminated. I depend on computers to exist." Chervin is a specialist, a man who builds models, bringing to bear his particular talent of understanding how data flows through a model. The mental world in which Chervin operates is, of necessity, laid out far more neatly than the physical sprawl he's built around himself.

Climate models sample points on their imaginary globes, they do not figure out all the equations that describe the weather for every place on earth. Three-dimensional models typically impose a grid over the planet's surface to fix their sampling rate. The equations of the model calculated at each grid point essentially average out the calculation for the space between points; it is assumed that conditions in Boulder will be basically the same as those in Denver, some 30 miles down the road, and so forth. In the NCAR version each grid point is actually 300 miles or so from the next, not 30—a typical scale for global models. The size of the grid involves a trade-off: more grid points gives greater accuracy, but each time the resolution of the model is increased by shortening the distance between each grid point, the amount of computer time needed to run the experiment leaps upward.

Whatever the scale of the model grid, however, events that occur in the 9,000 square miles or so between the grid points also have to be averaged in. This forces the scientist to make some more judgments, sometimes guesses. For instance, clouds, the most important of the so-called sub-grid scale phenomena, form and move on scales much smaller than 9,000 square miles, and the precise shape, composition and behavior of cloud cover is obviously important since wispy trails of cirrus will have a different meteorological impact than a system of thunderheads. Yet, Chervin acknowledges, "We don't really know how a cloud forms or all the environmental inputs that create and maintain a cloud. But we do know that the presence or absence of clouds can have an important impact on heat transfer within the atmosphere. So obviously to exclude clouds from the model is a wrong thing to do, so you have to do something."

Crucially, the model cannot *see* clouds and similar-sized events directly—they fall between grid points and hence cannot be derived from calculations within the model equations. The averaging of values derived from the large events that span the distance between grid points will not reproduce the effects of the smaller scale, and the resulting errors, as Edward Lorenz demonstrated with his simple model back in the sixties, will destroy the accuracy of the simulation. So the scientist must attempt to find a reasonable value for the impact of the sub-grid scale on the system the model is explicitly simulating and supply that number directly to the computer. For clouds, explains Chervin, the modeler usually tries to figure out the average amount of cloudiness and then see how the average cloud that his model contains affects the temperature of his model of the earth. If the results of the simulation match what is observed in the real world, the modeler is probably on the right track, but mistakes can still creep in. "You could have overly efficient precipitation, if you will," Chervin says, "and if you are producing too much in the way of cloudiness, your efficient rain could remove some of the excessive cloudiness—those sorts of compensating errors can produce the right answer for the wrong reasons." In other words, too many clouds inserted as a result of a modeler's gaff can be counterbalanced if the model happens to contain another failing, a function for rainfall that too swiftly consumes all the moisture in a storm system. The end number would be correct in that the model weather would resemble the weather of the real earth, but if the assumptions underlying the model are wrong, then the probability is very large that as the model runs through longer and more varied simulations, the errors will cease to cancel each other out, and the results gained from the exercise will be worthless.

The sub-grid scale problem derives from the fact that no model can ever be a perfect analogue of nature itself. The modeler's skill here is to paper over the imperfections in the analogy, to make the simplifying assumptions as accurate and as economical as possible. The fundamental barrier to gaining mastery over the model environment in which climate science occurs, though, is not in the gap between simulation and reality, for reality always offers a check to see how well the modeler is doing. Where the true difficulty always lies is within the model itself, in the task of learning precisely how the model itself behaves; the modeler must figure out whether the output is correct because the model is functioning well or because all the mistakes are, for the time being, canceling each other out.

This is why Chervin's "alpine approach" demands the greatest skill, for his solo efforts involve taking the basic model and modifying it, and every change may ripple through the behavior of the entire model, just as the flap of Lorenz's butterfly over the real Brazil can alter the weather two weeks later in Bahrain. The standard NCAR general circulation model is about 10,000 lines (or instructions) long, but like any substantial computer program (that is, virtually any program of more than a few dozen lines) the model is actually a hodgepodge of shorter programs strung together. A series of subroutines calculates heat transfers in the atmosphere, for example, and another group of shorter programs (using the results from, and feeding information to, the heat-transfer programs) calculates the motion of the winds. This mutual dependence between different parts of the program is what makes a three-dimensional model so difficult to use. It is crucial to guarantee that when one part of the model calls for a piece of data, it gets the right one, not a piece of information that has already been used by another part of the model and modified before it gets put back to its location in memory, and not a piece that has yet to be operated on by some part of the model that should already have gotten to it. It is in the movement of data from place to place that models most often fail by generating false results or by so lousing things up that the model overtaxes the Cray's memory, forcing Jensen to dump the whole package. It is Chervin's special talent, his special good luck, that he has a nose for where the data goes.

Today, Chervin lies on the border of exhaustion. He has just received a phone call from Cray Research in Minnesota; he turns to me and says, "We climbed Everest yesterday." He is too tired to show much more than relief, too tired just now to explain to me what he means. What has happened, it turns out, is that Chervin has been able to rewrite a model that runs on NCAR's Cray-1As and to run it on the next-generation Cray, a machine of radically different design that NCAR will begin to use within several months. A self-styled guerrilla scientist, Chervin has done it once again, slipping ahead of his colleagues to stake a claim on the capabilities of the new machine months before it arrives in Colorado.

This is where Chervin's world joins Jensen's. Chervin is adept at understanding how his familiar set of equations, his model, will behave within a particular silicon matrix, and his model has been customized to take the greatest advantage of that one type of computer; his thought, as captured in the model, has been molded to the shape of the machine that is, to use a nineteenth-century phrase, the engine of his thinking. Now the new and more powerful machine, with its different architecture, gives Chervin the chance to extract more information from his model since faster results yield more results, more detailed simulations, longer runs. But to gain this, Chervin must re-create his model world to conform to the structure of the new machine. The machine has forced on the model a new geography, a new map, and it is this that Chervin has had to master and that illustrates the novel mental processes that model-based science imposes upon disciplines used to the direct approach of laboratory-bench experimentation.

Chervin's special scientific interest lies with the question of what causes climates for a specific month to change from year to year. His research has involved running a model out for several years' worth of calculations. If over the years of the experiment the model generated certain patterns of climate but neglected others that have been observed in nature, Chervin reasons, then the climates that the model could not produce might be the result of some unusual event in the climate system, a boundary condition, that the model didn't account for. The trick would then be to find out what external event was responsible for the anomalous weather and climate pattern in the hope that it might then be possible to use this technique to predict whether a harsh growing season, for example, or a very warm, wet winter might result the following year.

The problem is that it takes a lot of model years and a healthy chunk of Cray time to be satisfied that you have come up with a sufficient range of weather. Chervin's version of the NCAR model calculates the weather at 1,920 points on the map; for each point, it measures the conditions at nine different heights in the atmosphere, which means that it must repeat its calculations nine times at each of the 1,920 grid points. It samples the weather every half an hour, going back to each spot on the grid and each altitude in the atmosphere; every twelve hours, the model tots up the day's averages and stores them on a tape that preserves this mock earth's history. Creating that history for each year takes ten full hours of a Cray's central processor time. Chervin's first experiment simulated the weather of twenty years. It took eight months of rolling in and out of the Crays to get the whole job done.

Eight months? It might as well be a lifetime. One school of thought at NCAR holds that if you have to wait more than eight hours on the pleasure of the Cray, you might as well take up knitting. Chervin puts it more mildly: "[Eight months] was at the far tail end of a normal scientist's attention span. One would like to get the results out sooner." Enter the new machine. The Cray-1A was state-of-the-art for almost ten years, and it was, of course, on that machine that Chervin ran his twenty-year simulation. But a decade is an enormously long time in computer development; the fact that the venerable Cray-1A remained preeminent for so long demonstrates how remarkable Seymour Cray's original break-through was. Finally, however, the first Cray has been surpassed by the Cray XMP, which Cray Research allowed Chervin to test-drive, and it is this device that has shaken Chervin loose from his scientific trap.

What makes the XMP special—what has provided the next leap in computational power—is its very special architecture. To steal a phrase from another discipline, the XMP is a house with many mansions. It consists of four interconnected computers packed into one box. Each of its four processors is almost half again as fast as the single processor in the Cray-1As. By bringing all four proces-sors to bear on a single job, the XMP can munch through calculations perhaps five times, possibly more, as fast as a Cray-1A. An XMP working on one problem at the Massachusetts Institute of Technology was clocked rounding the back stretch at the ungrasp-able rate of 1.1 billion calculations a second.

There is a price the scientist must pay, though, to gain control of the power the XMP holds out. As always, the structure of the new machine dictates the design of the scientist's thoughts. Compare the XMP with the older computers. The Cray-1A, like virtually all lesser machines, handles a problem more or less in the fashion that a person would think the matter through. It does one calculation, then the next, then the next. After it finishes all the arithmetic needed to figure out vertical wind speed, say at one point and at one altitude, it goes on to the next point or the next altitude. What distinguishes the Cray-1A is its speed, not any particular subtlety of approach or elegance in setting up a problem so that it can be solved in some spectacularly effective way. If programmed to canvas the weather on a mock earth from beginning to end in an order that does not send data winging its way to the wrong equations for the wrong locations at the wrong time, the Cray-1A will provide an answer in pretty good time.

Not so the XMP. Unlike the Cray and most people, it does four things at once. In fact, to make the new and expensive ($20 million dollars, with the bill going up as you add various bits and pieces that make the machine still more capable) computer worth its hire, you want it to use all four of its processors simultaneously for as many of the tasks in your program as possible; otherwise some very valuable computing power must sit idle, for your lack of skill in setting up the task in the first place. To make the computer work at full capacity, the programmer has to think like the machine: in parallel.

When Chervin works on a Cray-1A and must visualize what the world inside his model looks like, he doesn't picture a globe or a map spread out like the Mercator projections posted on a home room wall. He thinks in slabs of the model, that is, the results from all the grid points along a single line of latitude at each of the nine layers of his atmosphere. After the information on each slab of the weather is calculated and the results are stored, the model moves south, to the next line.

In principle, switching the model from the Cray-1A to the XMP ought to involve an obvious, almost trivial shift in perspective: instead of creating one slab of weather at a time, do four. That's simple enough to say, but it is considerably harder to bring about. With several processors working at once, you can't make any assumption about which processor will get to what part of the problem first. You must therefore create a program in which each processor can summon the data it needs when it needs it, plug that information into the equations for temperature, for example, and still be certain that none of the results from another processor's run through the same calculation has hung around to contaminate the next calculation. As Chervin worked through each of the subroutines in the NCAR model, he taught himself through a painful process how many ways data within a computer can get to the wrong place at the wrong time. "The easiest way to tell when you've made a mistake," says Chervin, "is when the model blows up."

During the final week before Chervin finally managed to get his new model in working order, the program blew twice. The first time he had forgotten both that the angle at which sunlight strikes the earth is going to be different for each of the four slabs on each pass through the computer and that that difference will alter the amount of heat each latitude of the earth receives from the sun. So he let each processor figure out the zenith angle for itself and store it for use in the thermodynamic equations within the model. But because any of the four processors might call on those equations at any time, the zenith angles were essentially getting picked at random, with just a one-in-four chance that the chosen zenith would be the right one for a given slab. That was one way to get garbage out of a very sophisticated machine, and it forced Chervin to write a routine that would wall off the calculation of the zenith angle in such a way that each processor would only recall the value calculated for the slab with which it was currently engaged. The other, and, Chervin hopes, last glitch came when he allowed each of the processors to calculate how high clouds rose over their particular slabs and then to store that value in the same slot in the XMP's memory, an error similar to that which created the zenith angle mix-up. "When the radiation routine came to use that value," Chervin says, "it was basically a crapshoot to see which one it got."

And then, after fixing those last two mistakes, Chervin reached Everest's top. He connived access to the XMP owned by Cray Research itself, and persuaded the company to let him slip his model in at night when the demands on the machine relaxed a little. Finally, on the day I met him, he received the call that the program had run itself out peacefully, duplicating the results he had achieved in a control run of the model through a single processor that had done its calculations in the old, familiar fashion. From the original model, 97 percent of the program could be run by more than one processor at a time; the XMP's extra power remained unexploited only 3 percent of the time. With one feat of programming, Chervin quintupled the science he can do. His twenty-year experiment will now require perhaps six weeks to roll in and out of the XMP; if he can muster the patience to wait another eight months, he can run his model out for a hundred years. The first experiment has whetted his appetite. He suspects now that seasonal variability is probably impossible to predict except under very special conditions, like those produced by El Niño, but until he can carry the model through for a century or more, he cannot test his thesis. Until he could get his hands on the XMP, he had no prospect of ever testing it.

With his considerable skill and his guerrilla sense of where the target lies, he has come up with a piece of work that allows him to learn today what he could not approach yesterday. That is one of the luxuries that the use of models provides for a man like Chervin.

Models are plastic, bendable to the modeler's whim. A creative scientist can take a mock world and ask questions of it that are new, receive answers that do surprise. Instead of science too large to be done except by some approximation of Henry Ford's assembly-line approach (like the great experiments in physics today that require the use of the largest particle accelerators, cost millions of dollars to perform, and require the services of any number of faceless, if highly trained, scientists to set up, perform, and interpret), suddenly a computer modeler today, once he has a model in hand and a machine on-line, can do his science on his own, at any time of the day or night, sitting out-of-the-way in an office in which he exercises his freedom to be messy. Chervin's model permits him to practice science as an individual act, with the questions he concerns himself with limited, at first pass, only by the bounds of his insight and the craft with which he manipulates his computer: that's one thread to draw out of Chervin's story—the essence of his alpine metaphor. But even Chervin, guerrilla scientist though he may be, is a machine-ridden man and as such is bound to a large and complex organization of devices and to the humans who maintain them.

A chicken-and-egg problem on a grand scale: does that science take place and advance because of the people involved or because of the apparatus with which they are surrounded? Both, of course. If Chervin does not wonder why one winter is different from another, the question doesn't get asked, at least not in his particular way and not at that precise time. From Chervin's perspective, the new science of climate derives its novelty from the fact that modeling has allowed him to escape from nature, in a sense; the malleability of mock worlds enables him to do much more, to think more freely, to create meaningful experiments that would simply be impossible in the real world. That novelty in its turn has been driven from the early days with Charney and von Neumann to the present, dominated today by the enormous power of the latest Crays, by the steady advance in computers and the techniques of their use. But for a final twist to the story of how modern climate science is done, we must take a further step.

At NCAR that last step takes us back into Jensen's realm, past the Crays and between the data-storage devices, through a door in the south wall of the room that houses the Crays. The doorway leads from the 1980s to the 1920s: bare concrete walls, dripping with leakage from plumbing grown weary. The Crays cannot run on ordinary 60-cycle electricity right off the wire. They forget their names, Jensen claims, in the valleys of a cycle that slow, so NCAR runs three motor generators continuously, 150 horses apiece turning out 400-cycle power, just for the supercomputers. The XMP will increase NCAR's power requirements beyond what the local utility can reliably provide, so Jensen is now building his own power substation.

The Crays turn their power into heat. Each of the 1,668 boards in the machines generates as much warmth as a 60-watt bulb. After five seconds without coolant, the solder that holds the chips together begins to melt; in two minutes the boards become irreparable. So each Cray is cooled by liquid freon, and each of the two machines at NCAR requires at least a 50-ton compressor to keep the freon flowing, with another 50-ton compressor driving the air-conditioning for the rest of the machines. Then there are the sources of power: a mass of 24-volt batteries that are stacked in a wire cage provide backup power for some of the machines; they save what's in the IBMs, for example, in case of a power drop. (When the power fails, the Crays must simply shut down—they consume too much electricity to run on batteries.)

Today a man stands atop a tall ladder over by the doorway as he pulls apart the plumbing that runs along the ceiling of the room; on the floor beneath him he has made a dank little lake. The only thing that's missing to complete the image of a plant at the heyday of the Industrial Revolution is some great boiler with a navvy feeding coal into the combustion chamber. The noise from the motors and the electrical devices is unbelievable. It's almost impossible to talk in the plant; in the computer room, computer operators of any seniority have lost the high-frequency end of their hearing to the persistent hum and whine from next door.

Chervin never has much call to leave his office and explore the computer center on which his (scientific) existence depends; he has nothing to do with the room behind it or its clumsy, noisy, unglamorous machines. These are the province of the plumbers and the electricians and the people who repair industrial motors. Everyone else worries about them only when repairs or modifications force Jensen to shut down his computers. But the context in which Chervin works and thinks is crucial in determining what it is he can do. Without the computers, without novel programs, without vast systems that gather data for the models, and without the closed rooms full of pumps and pipes and motors, the science of climate is hobbled.

Ideas are important, no question about it. But we must resist the image of ideas in a vacuum, ideas that appear full-blown, out of the brain of some brilliant scientist. The intellectual context, the culture in which a thinker finds himself, obviously shapes what he thinks about, the questions he asks, and the techniques he uses to answer them. Today it is impossible to conceive of climate science without placing the building and use of models at the center of the exercise, and the particular characteristics of how a climate scientist uses models, as Chervin's experience demonstrates, directs his research one way and not another. But the room full of motors and generators, pumps and plumbing that remains locked away behind NCAR's computer center points toward a larger context or culture that must also nurture the activity of people like Chervin if they are to thrive.

In other words, climate science is a big business. It requires enormous computers; airplanes to gather data; satellites to look down into the climate system; a vast network of people and devices just to keep the scientist and the machines he uses running at peak efficiency. What makes science advance? What makes the pursuit of climate science possible? What justifies calling the recent history of the field a "revolution in science"? It is the models and the creation of techniques that free the researcher sufficiently from the real world to permit his exploration of possible worlds, including the possible future of this one; the great systems of computers that, as never before, have made it possible to build and run sufficiently detailed models and generate results within the attention span of the average scientist; the individual scientists who have been able to make the imaginative leap from the habits of the laboratory bench into the novel realm of model and machine. Finally, it is the fact that the pursuit of apparently simple questions like "Can we predict tomorrow's weather today?" has culminated in the creation of its own infrastructure, a vast, expensive, complex set of institutions that our society has created, without which the machines could not function, the models could not run, and the scientists could not think.

And for Chervin, the context of science forms the context in which he leads his life, a line of tension that runs through that life. He walks me out of the building after our third meeting, after he has told the story of his climb up the hill and back again. After the awesome clutter of his office, this view at sunset from NCAR's porch seems to extend to an infinity of space. He tells me as he looks out across the edge of the Great Plains, "I'm truly blessed, in that I'm paid a reasonable (if not extravagant) salary to pursue my curiosity about how things work and have the opportunity to undertake these pursuits with very high-priced toys."

The sun is behind the Flatirons now, and out over Boulder the light falls in layers, cut by a scattering of clouds into a mixture of yellows, oranges, blues, and a deepening gray. Chervin and I watch for a while. It is quiet at the lab and the resident herd of deer remain off to our left, placid and undisturbed. Chervin speaks again. "I was leaving the building once, a year or two ago. It was early evening, at the same time of day when where I stood was in the shadows but the plains were all lit up. And for some reason or other the thought hit me that if whatever power there may be decides at this point in time that my time is up, and reaches down from on high and plucks me off the earth—I would not feel cheated. I don't think I was asking to be removed from the face of the planet, but I had a content feeling, that I've done A, B, C, which is really good; I haven't done X, Y, and Z, but that's O.K. That was as close to a spiritual experience as I've ever had." A pause. "And it probably holds true to this day. I'm not complaining."

CHAPTER **9**

Insurgents

IN HIS FILM *Sleeper*, Woody Allen plays a character who first lands at a house built of curves and glass that actually exists, perched on a hilltop between Denver and the Continental Divide. In *Sleeper*'s last scene, the action shifts to a laboratory where scientists are racing to clone a new dictator from cells taken from the old dictator's nose. That laboratory was NCAR, and one of the extras in those final, tumultuous sequences was a man named Stephen Schneider. In his daily life, Schneider is hardly used to filling an extra's role, and Allen noticed. "He told me," Schneider says, "he promised me that I would be in the scene, that my grandmother would get to see me. Then he told me *not* to look at the camera."

Schneider relates this story as we travel along I-70 west of Denver, heading for the ski slopes and a day away from NCAR. The story provides a good snapshot of the man, for within climate science Schneider has been more successful than in the movies at seizing and holding to the core of the matter. He has recently placed himself at the center of what has become the largest battle within the field, the contest over what climate science ought to try to accomplish next.

To a certain extent, the practice of climate science reached a plateau in the last several years with the development of sophisticated, three-dimensional climate models. With those tools, it became possible to make quite accurate weather predictions, and global climate simulations could be run out over many years. With more detailed models it has become possible to perform very complex experiments on particular pieces of the climate puzzle—tracing such practical problems as acid rain transport, for example, or the impact on the atmosphere, global weather, and climate of changes in the chemical composition of the atmosphere—as, for example, when carbon dioxide enters the air in unusual amounts.

But models are not the real world. Current models of global circulation are, at best, imperfect representations of one aspect of all the interactions among the elements of the climate machine; they describe the atmosphere more or less well, but the connections that bind the atmosphere to the oceans, to the biosphere, to the rest of the climate engine, are barely explored. There are absolute limits to what one can learn from these devices, these mock part-worlds, even when refined: while it is possible, for example, to detect a pattern of temperature changes that results from altering the carbon dioxide content of a model atmosphere, it is impossible, simply using the output from the general circulation model, to come to any conclusion about how such changes will affect the planet's plant cover. Lacking that knowledge, one can scarcely guess whether the patterns of water transport through the climate system will shift, and thus it becomes impossible to make with any great confidence a statement about just a single climate condition—rainfall—even when you know, because you have run the model yourself, precisely how another climate parameter—temperature—has altered.

Climate is not an idea that can persist for long in isolation. The modern science of climate began with the problem of predicting the weather; but the weather itself, and weather forecasting and weather-simulating models, only take you so far. The next great task the field faces, having developed credible models of the weather, is to devise a method of integrating them with the rest of the earth sciences—to build a broadly interdisciplinary science that is comprehensive enough to link the abstractions of a given model to the hard, material processes of the real world.

The first step in achieving that goal is obtaining enough of the right kind of information to feed the models. Models exist within a precarious mental space that is separate from the real world but dependent on an understanding of it. No model is any better than the accuracy of the initial conditions that it takes as its starting point, and those initial conditions are derived from measurements out here, on the ground. To reach beyond the study of the weather and into a science that encompasses the entire world puts an enormous demand on our ability to collect data. To meet that demand, climate scientists have turned to machines that are, after the computer itself, the greatest technological tools advancing the science: satellites.

The United States orbited its first satellites capable of remote sensing—gathering information from afar—in 1960 as part of the Discoverer series of launches. Those devices, which took photographs and then ejected the canisters of exposed film down toward a convenient ocean, were spy satellites. They supplanted the U-2 aircraft, which, following the destruction of Francis Gary Powers's U-2 plane over the Soviet Union on May 1, 1960, could no longer be risked over defended territory. Since that time the demands for accurate reconnaissance have pushed the development of ever more effective spy machines, from which have spun off various devices that climate scientists have been able to harness to their ends.

The images that the most modern satellites can acquire are striking. The space shuttle mounts a side-looking radar imaging system that can penetrate the cloud layer and produce extremely detailed ground-contour shots in a continuous strip all the way around the world. The most commonly used satellites, however, peer into other parts of the electromagnetic spectrum, taking pictures of the earth in visible light, or producing high-resolution images in infrared: taking snapshots, in essence, of the heat emitted by each spot on earth. Landsat, an American system, and SPOT, developed by the French, are both infrared imaging devices that convert their readings of heat emissions into numerical values and transmit those values to ground stations, providing an instant, or nearly instant, stream of pictures. National Weather Service satellites, though they scan a larger area than Landsat or SPOT and thus have lower resolution, possess four sets of sensors—one that can measure visible light and three that are sensitive to different infrared frequencies.

These systems permit the scientist to assemble a set of quantifiable, objective measures that can be compared precisely over time. Because each Landsat image begins as a stream of numbers, the values for two images of the same area, taken at different times, can be subtracted one from the other—with the result providing, for example, a very clear picture of how much forest has been destroyed or the extent to which a city has expanded. Satellites can likewise sense alterations in the flow of the Gulf Stream: in infrared images the gyrations of warm bodies of water that have been carried up the American coast from the tropics show up clearly against the background of the cold north Atlantic. The march of the Sahara into arid, semidesert regions can be seen in measurements that compare images in two of the frequencies sampled by the weather satellites.

Technology, again, drives science. While it is possible to model something that you cannot measure, the models will bear only a passing resemblance to the phenomenon of nature you may wish to investigate. Remote sensing permits the measurement of global events and provides the only system capable of documenting in a quantifiable way changes that occur on a global scale. And because of what the satellites can actually see—changes in plant cover; changes in the amount of water evaporating from a surface (evaporation involves heat exchange, which is detected by infrared imaging); certain measures of the chemical composition of the atmosphere; the success or failure of wheat crops in the Soviet Union; and so on—it has become feasible now, as it was not fifteen years ago, to conceive of a science that binds an understanding of climate with an inquiry into climate's impact on life and into the biosphere's impact back on climate.

A group at NASA's Goddard Space Flight Center has begun to develop a project with this kind of integration. With several partners, the group launched its pilot effort at the Konza prairie in Kansas, a tall grassland ecosystem. The program involves first wiring a small area of the prairie floor with measuring devices that monitor such parameters as soil moisture, the transpiration of water through plant leaves, and plant productivity. Next it examines the sites with very high-resolution remote sensing devices carried by aircraft and with more distant devices mounted on the satellite systems that overfly the region. The goal is to obtain through this network of measuring devices especially placed there a very precise picture of how that ecosystem behaves and then to use that picture to calibrate the satellite system. Once it becomes clear just how well the satellites can monitor the metabolism of the ecosystem, it will be possible to use satellite data within models that integrate the biology of a site with the observed weather. The

NASA group has already come up with a simplified model that does just that.

Despite such efforts as the Konza experiment, the central focus of most of climate research remains the narrower one that begins with the weather and moves away from it rather gingerly, step by step. Within NCAR itself the majority of researchers have dedicated themselves to acquiring virtuosity within a particular discipline: atmospheric chemists concentrate on the welter of reactions within the atmosphere; others study events such as storms and clouds that occur within the spaces between the grid points of three-dimensional models; atmospheric physicists, acid rain specialists, solar astronomers, and a large crew attend exclusively to the business of building each generation of the standard NCAR model. And the truth of the matter is that the choice to focus on an individual issue is obviously sound. For all the success of the atmospheric and climate sciences to date, the field is, in fact, just two generations or so old, and the amount that remains unknown is staggering.

Schneider himself, at the beginning of his career, began a study of clouds and cloudiness through which he demonstrated, within the models available in the early 1970s, that the special dynamics of cloud formation and structure could have a profound effect on temperature. He showed that increases or decreases in cloudiness due to a change in the carbon dioxide content of the atmosphere could play an absolutely unexpected role in the amount of temperature rise or fall that one would normally expect as a result of the new carbon dioxide level. The problem then and now is that the understanding of clouds is still too limited to permit anything more than educated guesses (or estimates, as they would more politely be called within the models) about what will actually happen to the cloud cover of the actual earth given actual changes in carbon dioxide or anything else. Clouds alone provide a fertile and crucial field of study for any number of researchers.

Despite his tightly focused beginning, Schneider, though he concedes that one person can't know it all, now argues that it's crucial to cross at least some of the disciplinary boundaries—at a minimum, to become capable of recognizing the important problems when they come up. Even more he wants to form groups of scientists into an institutional structure that allows a number of thinkers of disparate expertise to jointly produce some level of scientific understanding that can accommodate the irreducible complexity of the world outside.

Thus, in part what is occurring now in the field is a battle over turf: Schneider trying to wrest some intellectual ground, bodies, space, and money to pay for it all from established, ongoing work that Schneider himself would have to concede is worthwhile. Given odds thus stacked against him, Schneider has resorted to some organizational jujitsu. He and his colleague Starley Thompson have come up with a plan for a pilot effort to create an interdisciplinary research program that seems almost blushingly modest; Thompson has written the proposal in such a way that the first two years are going to cost only $150,000, a relatively trivial sum. That is enough to pay for some new lab equipment and enough computers to run a good geosciences data base, and the start-up costs for fourteen research projects in four areas: climate theory, paleoclimatology, the impact of trace gases in the atmosphere, and geophysiology—a neologism coined to describe the concept of inter-action between the atmosphere and the biosphere. As Thompson tells Schneider, the objective is to get NCAR rolling down this slippery slope before it knows what hit it.

"Once this gets finalized, the whole thing becomes the global systems division, already in place."

"That piece of sleaze," Schneider says, "is not going to slip by."

It isn't meant to slip by, says Thompson. "It's going to have to look like a division, smell like a division, and ideally it'll draw the best people already in the divisions we've got now, and take them into climate science."

It's a neat piece of strategic thinking on Thompson's part. Already the response of the existing divisions has been that they are already doing climate science and ought to get any new money that floats by. Should the new project emerge, the heads of established divisions could be tempted to cut their losses by trying to slough off their weakest scientists into what they regard as the fool's paradise of interdisciplinary work. If the project emerges after a couple of years as a viable new division, though, it will then have the institutional standing to ward off predators; says Thompson, "This isn't welfare for the weakest producers."

The imperatives in this battle to secure official standing in the organization are the same as those of the entire NCAR community a generation ago: without a place to nurture ideas, the ideas themselves may go begging in the wilderness. But an interdisciplinary organization must demonstrate that the questions it addresses are not only interesting but, more important, can be addressed by the tools of science, theory, experiment, modeling. Schneider's first priority is to win the battle at the level of scientific thought.

As a practical matter, Schneider is tenured and can do, more or less, the science that pleases him. He has too quick a mind to pursue only one problem at a time, so his research covers a lot of ground: he's trying to set up a conference to examine Jim Lovelock's Gaia hypothesis; he's finishing up a series of papers on the nuclear-winter issue; he's just started work on a couple of modeling projects designed to test ideas about the sequence of events that occur when an ice age ends; and last, but not least, he's running a range of model experiments to predict future climate changes caused by human activities.

This is the program of a restless man; partly what drives Schneider is the traditional passion of the scholar: nature presents itself whole, as a mystery, and the challenge of unraveling its secrets rouses scientific ardor. But politics comes into it, too, real politics, national decision making, not the battles of the lab. Schneider is convinced that we ought to be better prepared to anticipate climate change and the consequences it may impose on us; he views his work as a political act, using scientific advances to inform, perhaps mold political choices. The goal is to create a method of doing climate science that will endow its results with the power to persuade, to convince us outside the laboratory of the impact of a transformation in the way the world functions. We know that the climate alters—is altering—but what we lack so far is the ability to describe in detail what such novelty may mean to human life, day to day, year to year. Without that ability, there is little that the scientist may say that can inform the political choice that climate change may require. It is to gain a voice that Schneider, together with a growing number of his colleagues across this country and abroad, has begun what amounts to an insurrection within the climate science establishment.

The intellectual path Schneider is now taking is one of indirection; he is trying to demonstrate both the need for and the power of a new approach through an assault on a classical problem of climate science. With Starley Thompson, Schneider has proposed a reinterpretation of the history of the Younger Dryas, a period about 11,000 years ago when the warming following the last glacial maximum reversed sharply. The result was a sudden, brief return to the ice age: For several hundred years glaciers in Europe and North America resumed their advance. The conventional glaciologist's hypothesis about this reversal is that the warming after the ice age had proceeded to the point where the great Laurentide ice sheet (which covered much of the eastern portion of North America) and perhaps some of the European glaciers as well began to break up rapidly. This would produce a large flow of fresh water into the Atlantic; the fresher the ocean, the more easily surface waters freeze to form sea ice. And so the amount of ice covering the Atlantic would increase, spreading south; ice, being white, reflects sunlight efficiently; the Atlantic would absorb less heat, radiate away more; as a result, average temperatures would become, at least temporarily, low enough to produce the Younger Dryas glaciers on land. The process would reverse, the theory goes, as the global warming driven by the orbital effects and carbon dioxide transmitter continued to the point where the temperature increase was sufficient to melt enough surface ice, even on a relatively fresh ocean.

That account makes sense—it is a plausible sequence of events. What Thompson and Schneider decided to do, though, was to see if the idea worked in detail when tested within a three-dimensional model. After imposing a layer of sea ice on their model North Atlantic, they found, first, that the coldest land areas were near the coast and, second, that an ice-covered ocean affected temperatures on land most during the wintertime, not the summer. They concluded that during summers the most important factor determining how warm the earth gets is the amount of sun shining directly on land; the ocean acts as a heat reservoir to mitigate the effects of less sunlight reaching the hemisphere during the winter. Their model implied that the increase in sea ice imposed its cooling indirectly, by blocking the release of stored summer heat from the oceans during the Younger Dryas, as opposed to the conventional argument that sea ice worked directly, by reflecting solar energy back to space during the winter.

It's what came next that demonstrates what Schneider wants out of his field and of his colleagues. The impressive results achieved by Schneider and Thompson represent a creative and effective use of a model and develop that portion of climate theory that studies feedback mechanisms between land and ocean. Moreover, the results write up into a fine paper and dovetail into work being performed by others in the field. But that conventional bit of science, which took the two researchers a few months to set up, run, and write about, leaves unanswered the question of what actually happened in the Northern Hemisphere 11,000 years ago; what it was that set real glaciers in motion once again across the real earth.

To learn that requires an assault on the problem from at least two different angles. The first is to find some physical record of temperature change that preserves both summer and winter temperatures, with each season discernible as far back as 11,000 years ago. That evidence turns out to be fairly easily produced, though it requires the use of methods quite alien to a conventional climate theoretician. Schneider wants to look at beetles. Actually, he wants to get Russell Coope, a beetle expert from Britain, to dig up in Asia the fossil beetle record back to the time of the Younger Dryas. Since different species of beetles are sensitive to either the heat or the cold, counting which species of beetles are present at a particular location, charting changes in their abundance, and comparing tropical species with arctic species can produce a surprisingly precise record of maximum and minimum temperatures for any period for which there are beetles enough to add up. That kind of calculation is fairly easy to do; all it takes is the will to hire the scientists who can perform the measurements.

Even with that data, however, beetles cannot tell you *how* the ice sheets actually behaved. Resolving that issue demands a return to modeling, but not any modeling that can now be done. The difficulty here is one of the different scales of the events involved. The atmosphere and atmospheric models transform themselves extremely rapidly; oceans, wet or silicon, move more slowly, cycling over periods of weeks, months, and years instead of minutes, hours, and days. Glaciers travel to a rhythm of millennia. To model what could have happened to the ice during the Younger Dryas you would have to let all the models spin out for at least one thousand years. Running an atmospheric model that updates itself every half a model hour for a thousand simulated years is impossible with the machines now available. What's needed is a kind of model which does not exist yet—an asynchronous model—one that could run the atmosphere and ocean components sampling over years of weather and then extrapolate from that data the numbers it needs to make the ice calculations century by century.

In practice, computers large enough to handle even an asynchronous version of a fully coupled climate model are decades away, at least as Thompson sees it. So what was the rationale behind the two men's Younger Dryas experiment, which was, of technological necessity, an exercise in science that can be begun but not finished with the tools available now? For Schneider and Thompson, that experiment provides the exegesis of the text of their sermon: that there are limits on an individual discipline's ability to resolve the fundamental questions of how climate functions. Even the most intriguing answers are incomplete when produced from the approach of a single discipline. An atmospheric model with ice just slapped down can only tell you so much about how the earth warms or cools. The beetle angle, therefore, becomes a plea for better and more intelligent cooperation among disciplines that are barely on speaking terms.

And the vision of models unmakable for decades—for the careers of the people now proposing them? It's almost a joke Schneider is playing on himself. He has championed in recent years the argument that it's better to get a first answer quickly, using a simple model, than to putter around for years building a very complicated model before asking the first question. "First-rate science," he says, "doesn't require methodological macho. What matters is that it answers fundamental questions about nature. You need methods sophisticated enough to get you the answer, but essentially the methods are secondary—it's the answers that count." In terms of the original issues of what happened during the Younger Dryas, Schneider's approach resolves itself into the question of how does the earth, with its particular oceans and its individual continents, behave in detail when the average temperature changes. What happens in Paris, what happens at Minsk? Do oceans change abruptly or might sea level creep up or down slowly? Can we prepare for change, or will it catch us unawares?

The agenda is essentially a double one. First, Schneider wants to push himself and his field in a direction that will produce answers to the specific questions of how the world works. And in the world outside the Crays and beyond NCAR's walls, Schneider's goal is to mark out as clearly as he can the potential and the limits of what his science can do. What happens to Denver, what happens to New York should the world's temperature rise two or three degrees over the next fifty years? Schneider won't be able to tell you with certainty, or even with an assurance close to certainty; at best, by comparing one simulation with the next, he'll be able to give you a range of possible outcomes. So in terms of institutional politics, perhaps, his aim is to raise the money needed to pursue this kind of work by demonstrating what cannot be done now that might be very valuable to be able to do. And as for politics in the larger sense of the exercise of power within society, Schneider hopes that this work will compel the policymaker to recognize the extent and the limit past which specific results cannot serve as guides. The complexity that Schneider seeks to encompass in his science leads to the conclusion that any action we might take to prepare for climate change—whether it takes the form of warmings, coolings, changes in rainfall, changes in the amounts of acid falling to earth, the rise and fall of oceans, or the spread of ice or its retreat— anything that we might do we must do with imperfect knowledge, knowing that we gamble as we act.

It is dark by the time Schneider reaches this point in his story. We've skiied as much as our legs can stand, now we're on our way home. Schneider stops to call his wife, apologizing for leaving her to deal with both the kids and supper at the same time. With the perspective gained from an altitude of 12,000 feet up a mountain, his story—the progress from day to day, from paper to paper, in the ongoing battle at NCAR—proceeds in good order, with the ultimate outcome, the birth of the metadiscipline of coupled climate systems, seeming not in doubt.

What makes that victory appear likely, if not inevitable (in spite of the gaps in the accounts, the disappointments glossed over, and the failures now forgotten amid the catalogue of success), is that Schneider has carved a place in the vanguard of this revolution in science. He is the lineal descendant of the young Walter Orr Roberts, who through the windshield of a broken-down car conceived of a connection between the physics of the sun and the desiccation of the fields he passed. Schneider is the heir to the triumphs of method that computer designers and model builders have gained in their attempts to re-create the behavior of different parts of the climate system.

In point of fact, Schneider's predecessors have probably done most of the work. But what distinguishes climate science today from the studies performed by any of the atmospheric scientists, oceanographers, or geologists of a generation or two ago lies in the fact that in the years after the war people such as Edward Lorenz and Philip Thompson and the rest learned how to do the mathematics needed to model a nonlinear system like an atmosphere. Combine that mathematical skill with the decade-by-decade leap in computer power, and you have what Schneider needs to pursue his particular agenda.

As for Schneider's own place in all this—well, to carry along a martial metaphor, Schneider is playing Kubilai to a preceding generation of Ghengis's, or perhaps more complimentarily, George Washington to a band of Thomas Paines. Whether Schneider wins his particular battle at NCAR and gets the money within that laboratory to mount his multidisciplinary attack, his is clearly the direction in which the science as a whole will go.

There is a practical reason why that should be so. The world we live in today is a product of climate history extending back to the origin of the earth; it is a system imposed upon us by the particular accidents of time and place that went into the making of this earth. Until recently the study of the earth and its climate tried either to write that history or to guess at the very near future, predicting tomorrow's sunshine or early morning fog. Now, for the first time, we have the potential to modify our climate on a large, possibly global scale; where climate changes used simply to happen to us, we are now imposing them on ourselves. What's more, such changes are, for the most part, without precedent. It is impossible to look back into the physical record of climate, back into ice cores or the strata of ancient rock, to find a direct, unequivocal record that can, with any certainty and detail, map out the consequences of the processes that human society in the last century has set in motion. We are altering the world in ways that will cost us, that is certain. But to understand properly the meaning of a human-induced leap in the carbon content of the atmosphere, or an alteration in the reflectivity of the earth's surface caused by choices in agriculture or animal husbandry, or the possible range of consequences for forest cover that accompanies air pollution, we must embrace the tools of climate science that let us explore the variety of outcomes that may await us. And to persuade us that the model results have some bearing on life within the real world, it is necessary to integrate climate science with a perspective that embraces the weather with changing conditions over time, with the life cycle of plants and of ecosystems, with changing human demographics, and ultimately with the pressure of human choices that will be made, with or without knowledge. That's the spur, the goad that is driving the field Schneider's way: the need to get some sense of what the range of possible futures may be.

Wallace Stevens's poem "Variations on a Summer Day" contains this verse: "This cloudy world, by aid of land and sea/Night and day, wind and quiet, produces/More nights, more days, more clouds, more worlds." More worlds: Schneider's stock-in-trade. This night, after this sunlit day, Schneider invites me to his home. His wife, Cheryl, has finished making supper, and their two children are almost ready to be stowed out of sight. We drink wine and talk, and Schneider returns to the message with which he began, as if trying to make certain that if I remember nothing else, I will understand this. The idea behind what he does, Schneider says, isn't just to make the models behave. Rather, if he gets the Younger Dryas right, he'll be more ready to believe that he can predict what will happen as human beings add methane, for example, or nitrous oxide to the atmosphere. The actual ray of sunlight absorbed by an actual molecule of methane is what counts, not the number of computer instructions it took to predict it. We talk, and drink, and move on to other subjects until it grows very late and I head home. At the end of that evening Stevens's imagery lingers: such nights, such days, the clouds within the computer, its winds and lands and seas produce many worlds. With each day, more worlds, and from among them, with the choice of human actions, with the choice of sciences to study, comes the power to make over one world, this cloudy world that we inhabit.

PART III

THE CONTAGIOUS DESERT

FROM climate as it is to climate as we imagine it to be, inside the machine, take the last logical step to climate as it will be, the future of the real world out there made knowable, in part, by spinning out possible futures within the electronic worlds.

Climate science has developed as an effort to understand how the world functions. It advances by producing a more coherent picture of the historical processes that led from the beginning to today. The measure of its success is how well the worlds in the machine approximate what it is we can actually measure out there, in the real world. We validate our picture of reality by comparing it, always, with what has just passed.

But the story and the science now turn and from our present peer, not backward, but forward into our future. We ask, with an increasing sense of urgency, what will happen next? The power of models derives in part from the fact that they don't notice what time it is. Whether you start them up with today's measurements or Thursday's, they will happily spin out some image of what the world might be like next week, next year, or, if you can wait long enough, indefinitely.

All such efforts involve imagination, some willingness to speculate, to compose stories that are not quite fictions but that nonetheless are not scientific fact in the familiar sense of a measurement of some real quantity we can find in a rock stratum, at the ocean floor, or in the stratosphere. As statements about the possible range of events that may transpire, simulations of our unfolding climate are quite real; and as models grow in power and resolution, as theory advances, and as the pace of change actually occurring in the world upon which we live accelerates, these tales of our future become ever more plausible and ever more frightening.

The new science of climate has developed a picture of the climate system that begins with the most general phenomena and builds upon them to reach the most particular—moving, for example, from the creation of an atmosphere over the longest sweeps of time to the swift and local fury of a storm, the weather that is the most accessible piece of the climate machine, the weather that rains or shines on us each day.

Human-induced climate changes follow the reverse track. The most easily understood and most obvious consequences of human action come with changes in the weather we now experience, local events that pass quickly or that are confined to one place or region. That observation is not new: cities have long been recognized to alter the climate of their immediate environs, usually being a degree or two warmer than the surrounding outlying regions. Such a temperature shift is, after all, a simple form of climate modification. But what is new is that the human influence on the weather is being sustained long enough and with sufficient breadth so that climate changes are occurring on a global scale, changes that will take decades and centuries to work through.

So the next four chapters move from the particular (acid-laden storms that strike one hillside at a time) to the somewhat general (regional shifts in climate that can lead to the creation of new deserts) and finally to climate changes that are universal, that alter the relationship between life and the weather worldwide for the indefinite future. Different processes are involved at each scale of space and time, but the common thread is this: The deeds of human society are transforming climate.

What follows, then, is a catalogue of disasters, some already in the making, some possible, and one that is, I hope, nothing more than a fairy tale, a grim and monitory fable. All of the stories share that common theme of the leap in the ability of human societies to affect climate. Now our powers to dissect the climate system are striving to keep pace with a rate of change in the climate that is driven by active human intervention.

Paradoxically, that active agency gives hope in the face of the great risk that climate transformation poses; what we do we can understand, and what we come to understand we can, if we choose, alter. Human reason, in the form of a new science of climate, is now able to provide a certain kind of picture of the world that we are creating. The issue now is whether our ability to recognize the implications of what we do will alter in any way the decisions we must now make in the face of the dangers our actions so far have created.

CHAPTER 10

When People Make the Rain

I N THE WHITE MOUNTAINS of New Hampshire, there is a small village called West Thornton. It sits in a valley, just alongside I-93, the interstate that runs from Boston up through the ski country. The Pemigewasset River runs through the village, heading toward the Merrimack; the mountains rise on either side of the river valley, and on clear days, some miles to the northeast, Franconia Notch stands out. Off to the left, when you face north, Hubbard Brook joins the Pemigewasset. Hubbard Brook itself rises out of the confluence of several small headwater streams draining a bowl of hillsides up to the ridgeline just west of the village; the entire network of tributaries and branch streams drains over 3,000 hectares of forested ground. It is, as near as may be, an absolutely typical patch of land for northern New England. The Hubbard Brook watershed does have one distinguishing feature, though: a tract of land formally known as the Hubbard Brook Experimental Forest, which is probably the most intensely studied forested ecosystem in the world.

In early August 1984, the weather over the eastern United States was quite ordinary. A high-pressure system sat stolidly in place for about four days in the first week of the month, and as days go they were pretty doggy: hot, muggy, and slow. On the seventh a mild low-pressure zone started moving north to south; winds at the surface began to carry air out of the Ohio River valley toward the East Coast; when a cold front stalled off New England, much of the region was treated to a classic summer rainstorm. Over the next few days, the winds reversed themselves and began to flow more or less on the diagonal, from northeast to southwest, carrying clouds and fog halfway down the eastern seaboard.

The storm and the days of gloomy overcast that followed passed over the Pemigewasset valley; West Thornton and the Hubbard Brook watershed got rained on, then socked in. On the twelfth of August, technicians monitoring the research forest's instruments collected a sample of cloud water that had accumulated between 6 A.M. and noon for pH testing. (A pH test measures the acidity or alkalinity of a substance. The lower the number, the more acidic the substance.) When analyzed, the sample turned out to have a pH value of 2.97; that is, condensing water within the clouds and fog was about 10,000 times more acid than clean rain. Along the coast, at Bar Harbor, Maine, samples collected from the seventh to the ninth all showed pH values of 3.00 or below. In the Shawangunk Mountains near New Paltz in the Hudson River valley, the pH dropped as low as 2.80, and even as far south as Loft Mountain in Virginia, the clouds still contained enough acid to produce samples with pH values just a hair above the 3.0 mark.

An obvious conclusion to be drawn from that one week in August 1984, or from all the other storms that have come and gone over the past few years, is that as a matter of routine, as a purely ordinary feature of the weather, the entire Eastern seaboard must get bludgeoned periodically by intense bursts of extraordinarily acid pollution. Certainly the record is clear enough; what it means is not. Beneath the question of whether or not acid rain matters is a consideration of vision and beliefs—what your particular outlook permits you to examine and decide. Although couched in the language of science and circumscribed by its methods, at its heart the issue is not a scientific dispute. I believe that acid rain in particular and air pollution more generally threaten the natural world by repeatedly exposing it to climate events that are, in essence, man-made. I argue that the consequences of such exposure decrease the resilience of the planet upon which we live and that continuing the practices that produce the pollutants and the acids puts us at ever greater risk. Most simply, I am afraid of the damage that acid rain may be causing, and I hold that we should put a rein on the actions that cause it.

What follows here is advocacy, an attempt to persuade others of my opinion. Begin with a metaphor for the entire acid rain issue: the amazing smokestack built by the International Nickel Company at Sudbury, Ontario. It is one of the great architectural monuments of our age—a superstack more than 1,000 feet high used to vent the company's smelter. At its worst, that stack poured out 5,000 tons of sulfur dioxide each day. Although it is finally a problem of nations, of collective totals of burnt coal and oil, of winds that travel without respect for borders, in principle, at least, acid rain begins with single, specific increments of sulfur and nitrogen, and it ends up as part of a particular fog bank or mass of haze on an uncloudy day. So take Sudbury as a symbol, an archetype, as simply the most fantastic of many possible starting points.

The whole rationale for the tall stack is that it injects wastes high enough into the atmosphere to minimize the pollution in the already devastated lands right around the plant. What actually happens is that uncounted molecules of sulfur dioxide enter the atmosphere high enough so that some large proportion of them will be lofted by vertical winds up to levels where they can reside for an often considerable length of time. There the combustion products encounter a variety of compounds: ozone, hydrogen peroxide, ordinary water vapor, peroxyacetic acid, dozens of different chemical species. What occurs next depends on just which compounds each particular molecule of sulfur dioxide encounters, but in one of the simplest chains of events the sulfur dioxide molecule becomes, after several intermediate steps, a molecule of sulfur trioxide (SO_3), which then combines with a molecule of water (H_2O) to produce a single molecule of sulfuric acid (H_2SO_4).

Imagine, for the sake of the argument, that all this is happening in early summer. By author's fiat, I place a high-pressure zone over the midwest of Canada and the northern United States. An inversion has been in place for a few days, so it is getting pretty hazy over Sudbury; with the winds blowing slowly toward the south and east, the haze and thick, muggy air spread, too. Despite the height of the stack, some of the sulfur from Sudbury settles near the plant, falling to earth attached to specks of dust, but some of it remains airborne, entrained in the moving columns of wind. The plume from a tall stack can spread, even without extraordinary atmospheric conditions, along a widening track hundreds of miles long; the ability of the atmosphere to transport pollutants literally around the world was demonstrated most recently by the grim drama that played itself out as each succeeding nation waited for word that the cloud of radioactive fallout from the burning Chernobyl reactor had finally reached its territory.

So, again, by a not altogether arbitrary fiat, let the breakup of the high-pressure zone create a wind pattern that carries some of Sudbury's sulfur east, toward the Atlantic coast. Then, as is common enough in the summer, a few thunderstorms develop in the neighborhood of the Connecticut River valley and begin moving east on their own. The boomers produce spectacular rainfalls, the kind that give you an inch an hour as long as they last and flood all the roads that have settled a bit more than they ought. Finally, a thousand miles and more from the Sudbury stack, one raindrop strikes one leaf on a birch tree on the banks of the Pemigewasset River, a raindrop tainted with a leaven of sulfuric acid.

For Sudbury and New Hampshire, read Ohio and Nova Scotia, or northern Mexico, Arizona, and southern California, or York, Stockholm, East Germany, and Estonia. The point is that acid precipitation can occur whenever combustion occurs—and what goes up as smoke can, and does, come down half a continent away. Examining the microcosm—one tract of woods in a sparsely populated region of the United States—illuminates the scientific issues that have emerged within the international political debate about acid rain, and the single location stands in for all the land upon which acid rain now falls. If you can measure any sulfuric or nitric acid in your rainfall, that acid almost certainly was produced as a result of human actions. More formally, in a spring 1986 report on acid deposition issued by the National Academy of Sciences, the Academy panel concluded that "in eastern North America a causal relationship exists between anthropogenic sources of emissions of sulfur dioxide and the presence of sulfate aerosol, reduced visibility, and wet deposition of sulfate." That is, there is a direct connection between the pollution that goes up and that which comes down.

In the eastern half of the United States, somewhere between 10 and 15 million tons of sulfur get poured into the atmosphere each year; for oxides of nitrogen the possible range is wider, with emissions ranging from 10 to 20 million tons annually. The burning of fossil fuels produces over 90 percent of the sulfur and nitrogen that returns to earth as sulfuric and nitric acids. About two-thirds of the acidity in acid rain in the East derives from sulfur, the rest from nitrogen; 70 percent of the sulfur and 30 percent of the nitrogen comes from electric power plants and smelters. Transportation, essentially car exhaust, accounts for another 40 percent of the nitrogen. That's it—if you want to know what causes acid rain as a nationally averaged phenomenon, these are, in essence, all the facts.

Our knowledge is no less refined at the next levels of detail. Atmospheric chemists basically agree on the reactions that produce the two acids out of the original compounds; although several other chemical pathways exist, in addition to the one described above, the basic ideas are set. The results of that chemistry, the acidity in rainfall, are easy to measure, and, as in the acid deposition event of August 1984, it is even possible, by dint of a reasonable amount of logistical maneuvering, to map the extent and intensity of a single acid storm. Even the impact of acid rain on certain kinds of ecosystems is beyond question. Acid rain has acidified some lakes; the exact number within the United States is not known, but over two hundred lakes in the Adirondacks are so acidified that they can no longer support fish populations. Acid deposition can destroy stone monuments, and acid rain has been implicated in the damage done to the U.S. Capitol. Acid rain also certainly has some impact on flora; in the laboratory, for example, acid mists have damaged leaves and slowed plant growth. For a real-world example of the hazards it poses to plant life, most scientists involved in the question now agree that acid rain is a major, if not the sole, culprit responsible for the significant damage to the red spruce forest at Camels Hump, a peak in the Green Mountains of Vermont.

Although the harm done to the Vermont forest is perhaps the most unsettling legacy of acid rain, forestry problems have been observed throughout the eastern woods. A number of tree diseases have been reported with increasing incidence. A study at the Oak Ridge National Laboratory, which examined more than thirty species of trees from the eastern forests, showed evidence of retarded tree growth. In the higher altitudes of the Great Smoky Mountains in North Carolina, groves of trees have simply died. Clearly something is weakening a number of American forests.

Place that in a worldwide context: Germany's Black Forest has been the scene of one of the most devastating diebacks observed, and other forests within that country are showing signs of severe damage. The destruction there has been linked to air pollution, including acid rain. Marble statues in Venice have been dramatically eroded by the acid generated by the intense pollution within that clogged city. As many as 20,000 lakes in Sweden have been acidified.

The United States does not currently suffer from as severe an ecological problem as those observed elsewhere (though particular locations, like those bathed in acid fog generated by the concentration of car exhausts in the Los Angeles area, or those in the immediate neighborhood of large smelters, can show dramatic effects). But the disasters suffered abroad couch the debate that America now faces. We may simply be luckier than the Germans, possessing forests more resilient, soils better able to buffer acid effects, and levels of pollution sufficiently lower to avoid the worst impact of acid rain. Alternately, the experience of others may be a warning and a guide, a harbinger of what we might encounter if we proceed along a proven path. Guess wrong one way and we spend a lot of money to no good end; guess wrong the other way and we face the prospect, at worst, of losing the natural landscape of the eastern United States.

I know on which side I would choose to err; it is unacceptable to me to gamble, at the odds we seem to face, with so much of our natural habitat. Whatever we do about the problem of air pollution, there is always the chance that we are taking the wrong steps, and nothing I can say here will eliminate that chance. Uncertainty, the absence of clear-cut scientific conclusions, especially on the question of what is the actual, present danger that faces American forests, is intrinsic to the issue; science, the chemistry of acid generation, the biology of acid effects, can only outline the risk, not detail with absolute confidence what shape the future will take. Thus, once the basic outline is established, acid rain must become a question of politics, of deciding whether, how, and how much to control the emissions that cause the damage.

Politics, not science. A choice must be made between the dollars a control program would cost and the potential benefits of reducing the acid content of rain and dry fallout, that is, the man-made component of the weather. It is an ordinary enough political problem, and in the ordinary fashion partisans of controls attempted, at the time the decision was being made, to lobby the man in charge. In September 1983, William Ruckelshaus, then the administrator of the Environmental Protection Agency, took a group of five scientists to the White House to try and talk President Reagan into backing the EPA's plan to reduce sulfur emissions nationwide. As one of those scientists, Gene Likens, director of The New York Botanical Garden's Institute of Ecosystem Studies, recalls, "We told the President that the overall effect of acid rain is very clear." It was a fairly cordial meeting. Reagan's first interior secretary, James Watt, passed around the famous jelly beans, and near the end of the discussion, Likens says, "The President, trying to be cute, I assume, said his liberal arts education had not really prepared him for such complicated issues as this. Everyone chuckled."

Complicated or not, in subsequent years Canada and, as of this writing, six American states established programs to limit sulfur pollution within their borders. The United States as a whole, however, has done nothing. Ruckelshaus was for a time the point man, defending the administration's decision not to act and claiming, in public at least, that the sole reason for the delay was the inadequate base of scientific knowledge about acid rain. In particular, he argued that the best researchers in the field could not tell him, a policymaker, how severe the damage was that acid rain might be causing, how quickly that damage might be getting worse, or even how many years it would take to find out. Under the circumstances, he confessed, his hands were tied: how could any rational politician make a decision based on such imperfect information? Ruckelshaus made clear in his testimony before the Senate who was to blame for the holdup, and hence for any continuing pollution: "I am sure," he commiserated, "that you share with me in feeling a certain frustration when it comes to the timeliness of scientific results."

It is August 1986. Gene Likens spends each of his summers at Hubbard Brook, and he has invited me to his home, a rented condominium that stands on cleared land on the east bank of the Pemigewasset, a couple of miles downstream from the confluence of that river and Hubbard Brook. With us is one of Likens's closest collaborators, Herbert "Fritz" Bormann of Yale University. By way of introduction to the subject at hand, Bormann points out the picture window that faces the solid forest wall on the other bank of the river, and says, "You look over there, and you see a forest that is doing fine—it's green and tall. But suppose it looks great; productivity in the forest could be down 10 to 15 percent and it would still look good. It could be, right over there, that there is a substantial reduction in the growth of that forest, but one that you could never see." It is a forest bathed regularly by acid rain and fog, yet Bormann, an ecologist, cannot say what, if any, damage the acid has done to the forest he knows best; like me as I look at those woods for the first time, he knows that something is happening within them, but he does not with certainty know what it is.

In fact, Ruckelshaus was right when he testified in Congress and is still, in broad outline, right today: science (Likens, Bormann, and their colleagues across the country) cannot provide him with the answer to the question of what, precisely, acid rain does to the environment as a whole. But there will always be an absolute conflict between what the politician wants, or says he wants, and the goals of the science in question. Politically speaking, a firm statement to the effect that acid rain causes x, y, or z dollars' worth of damage to the forests of the Northeast might make the EPA's life a little easier. The only difficulty is that that kind of assertion cannot be made: there is no way to tell, before controls are imposed or acid rain has had enough time to do all the damage it can do, what the final result of any choice of action in this area will be. Even if that kind of perfect knowledge is truly what Ruckelshaus sought, he could not have hoped to get it.

Actually, as a purely scientific problem leaving aside any considerations of politics, the acid rain issue has a kind of elegance as a model for a class of problems that casts into relief the gap between the usual disciplines of science and the irreducible complexities of the climate system. The "easy" (in relative terms, of course) end of the science is the sort of research that can be performed in the classic manner. Measuring emissions, modeling, tracing chemical synthesis, analyzing rainfall—all these, though complex individual problems, are straightforward enough conceptually.

The science gets harder as soon as the question arises of tracing the links between the emissions source and the place where the acid ultimately lands. In my schematic of the process, the route was Sudbury to New Hampshire, a connection that almost certainly occurs; but since acid rain is a function of weather, and that raises the issue of Edward Lorenz's butterfly, there is a limit to the predictability of the atmosphere, which sets some absolute bounds on our ability to say whether a particular plume of sulfur will produce a particular increment of acid on a distant hillside.

And finally, when that acid lands along the Pemigewasset River, the simple question of what happens becomes virtually unanswerable. It is possible to look at the landscape around the Sudbury stack, and conclude that the devastation, the really miserable state of the vegetation there, is almost certainly caused by the presence of a single, overwhelming stress—the pollution from all the low stacks that preceded the superstack. That's really no more difficult than suggesting that a spray of defoliant is probably responsible for knocking the leaves off your mulberry bushes. But come as far away as Hubbard Brook, and the intractably dense weave of interconnections that dominates any piece of landscape creates a fundamental intellectual obstacle for anyone trying to understand it. Walk up the hill with Fritz Bormann, and it becomes clear how difficult is the simplest question; walk up the hill and find the immense territory of the unknown within any single patch of land.

Up one of the ridges within the experimental forest is a creek, one of many flowing down this particular face. It has its own watershed, fed first by the rain that falls on that particular piece of ridge and then, as it descends toward Hubbard Brook, by its own little network of tributaries, stream, branch, and trickle. A dirt road lies on a cut well above the creek bed; that way the soil loosened by the passing traffic will simply settle into the hillside instead of clogging up the waterway. The thick woods thereabouts are a mix of red spruce, sugar maple, beech, and yellow birch. About halfway up the ridge the road levels off for a moment before opening out to the right into a cleared grassy square, about fifty yards on each side. Another hundred yards up the hill and suddenly there are no more woods. The entire watershed, from that point to the crest of the ridge, has been logged out, with the trees being taken first to the vacant lot fifty yards below, and then out for sale.

Bormann is a man of middle size, with some gristle on his frame, who looks like someone who has spent a great deal of his life out of doors. He and Gene Likens have been walking over the ridges of the Hubbard Brook Forest for almost twenty-five years now. Bormann has a game foot today and so pauses from time to time as we hike upward, but in between moments he seems to forget the injury, scooting upward fast enough to make me scramble, until some twinge reminds him that he is not altogether fit. Just at the edge of the clear-cut, he pauses to point out a trunk that rises straight as a die—an ash tree. "These are one of my favorites," he says. "They make baseball bats out of these, which is very satisfying to me." Twenty feet farther up the hill the forest ends.

In its place is a jungle, shrunk down to about chest height. There are a fair amount of pin cherry, a plant that buries its seeds and sprouts only when a disturbance occurs that gives it light and room enough to grow; some moosewood sprouts; and the first tiny seedlings of the trees that will presumably reclaim this turf for their own. Bormann bends over to find a yellow birch three inches tall. Planning for the clearing of this area began three years before, when a phalanx of graduate students spread out across the hillside to count everything they could find. They laid out a grid and then mapped every tree more than four inches in diameter. The U.S. Forest Service built a monitoring station on the creek just below the area to be cut, and then the research group called in a commercial logger to take the whole forest down.

And then they waited to see how the ecosystem as a whole would respond—where the land becomes vulnerable and what form the damage control takes as the forest returns to the clear-cut area. For example, Bormann has measured the flow of nutrients that had been trapped in the forest floor but are now escaping out into the creek; before the cutting some researchers analyzed the soil. Researchers are doing it again this year, to check the development of new topsoil. "At first the stream shows very high nitrate levels," implying the rapid loss of nutrients from soil unprotected by forest cover, says Bormann, "but it's interesting—the stream comes under fairly rapid control of returning vegetation—maybe three to four years."

Bormann leads me up, over one of the side streams and then along the edge of the clear-cut area. He stops by a hay-scented fern, a dense clone that has seized a small plot and largely crowded out all other growth. Bormann pauses again, using each delay forced on him by his injury to point out another detail of the accumulating picture of the forest, and says, "This is what's fun, getting to see how this all works, seeing all the systems in action."

In this one spot, on this one afternoon, all the systems seem intact. The three-year growth is virtually impenetrable; the streams are running clear. Even with the immense disturbance wrought by a clear-cut logging operation, what Bormann shows me is obviously still a vigorous, vital patch of land. When the experiment is complete, that is, when the original forest seems well on its way to reclaiming its turf, the whole exercise will have answered questions on a variety of matters: What are the comparative nutrient budgets of a healthy forest and of a disturbed one? How does a forest regulate the hydrology of the stream system that drains it? What, in ecologists' terms, controls the rate of productivity (that is, new growth)? But, and at last, the long way back home, what neither Bormann nor any of his colleagues can do is to take that piece of land, watch its regrowth, and identify any single factor that speeds or slows the process. There is, in fact, no clear way to determine whether the forest is flourishing or merely surviving; researchers will not be able to say whether a three-inch-tall beech seedling is abnormally small or stunningly robust after these three years of growth. "As scientists," Bormann explains, "we are taught to be experimentalists, taught to vary one or two aspects of the environment for your organisms. For acid rain we can do that in the lab—we can mist a spruce seedling with some acid. But here in the field I have no way of separating out the effects of acid rain from anything else that may be happening to the forest."

Bormann's argument hearkens back to the question that prompted our walk up the hill: What happens when acid rain strikes Hubbard Brook, or anywhere? The honest answer is "I don't know," at least not in detail. The forest is not a system in which events occur in a nice, neat linear chain of cause and effect. Plenty of things are happening. Ozone strikes the trees, and certainly damages them. Days of sun and days of frost leave their own traces, each day adding its increment of effect. Pests come and go. Lead levels used to be high in rainwater but have dropped now that unleaded gasoline is common. An old tree dies and topples, leaving a swath of light and air cut into the forest canopy, an opportunity for shadowed plants to shine. Each of these events has its impact, all in a natural reprise of Bogie's famous line, "Maybe not today, maybe not tomorrow, but soon and for the rest of your life." But Bormann cannot tell me, nor I you, what does what to whom, and when.

In the real world different pollutants often strike at the same time. Not only is it impossible to separate one poison's effect from the next, but lab studies have been able to show that pollutants working in concert can either magnify the damage they cause individually, leave it just about the same as if the various compounds had hit the ecosystem in succession, or even, in some cases, mitigate the effect of the overall load of wastes. In a place like Hubbard Brook, the response of the watershed that Bormann showed me to a dose of pollutants could be—actually, almost certainly must be—somewhat different from the reaction of another watershed 1,500 feet higher; that is, the same combination of sulfuric acid, ozone, nitric acid, cadmium, lead, sunlight, shadow, radon, cigarette ash, and whatever else might come that way could do great harm to one stand of trees, but do much, much less to another a few hundred feet of elevation away.

Simply, the best a scientist can say and remain honestly a scientist is that acid rain is certainly bad for several species of plants that have been tested in the laboratory, including many of the major species that dominate the forests of the eastern United States. He can say that some damage to the natural ecosystem, occasionally severe damage, has been seen in areas that routinely suffer significant acid pollution. He can say, as Bormann does and Likens did, to President Reagan, that reducing the amount of sulfuric acid and sulfate that lands on the eastern United States is probably a good idea. What he cannot say is that acid rain is the direct and unequivocal cause of major damage to the eastern forest and that a program of control will save ecosystems worth some specific number of dollars.

It is that sudden lapse into silence that really creates the argument about whether we must control acid rain. David Stockman, more blunt than Ruckelshaus, put the issue into terms easy enough even for Reagan, with his liberal arts education, to understand. In a speech made as far back as 1980, when he was director-designate of the budget, he laid down the criteria he would use to analyze the controversy:

> I kept reading these stories that there are 170 lakes dead in New York that will no longer carry fish or aquatic life, and it occurred to me to ask the question . . . how much are the fish worth in these 170 lakes that account for 4 percent of the lake area of New York? And does it make sense to spend billions of dollars controlling emissions from sources in Ohio and elsewhere if you're talking about very marginal volume of dollar value, either in recreational terms or commercial terms?

Problems of vision, of belief, of what you can see and what your beliefs permit you to see. If Stockman were sincere, then he may honestly have been blinded by his beliefs; if Ruckelshaus had faith in what he told the Senate, then he is a man deluded in his understanding of science. The measures they ask of and the standards they set for science are false tests. By seeing the question as one of economics, Stockman missed the point: A price tag cannot be placed on a complicated system, since you cannot at any one moment identify precisely what's gained or lost. Ruckelshaus craves clarity and a clear chain of cause and effect, certainty, from science. And yet, again, until the events have transpired and the past is irreclaimable, the natural world itself will impose a limit on the certainty with which Ruckelshaus's erring scientists can advise him. Through their eyes, through the eyes of the Reagan administration, acid rain is not a problem until it makes itself apparent unequivocally, until it can be measured as dollars lost to the till or acres lost to the blight. It is a plausible vision, perhaps, but not one justified or demanded by the state of the science; it is simply a choice, a political choice.

And it is a gamble, pure and simple. To acquiesce in the Stockman line of thought, we have to hope that acid rain turns out to be a minor problem, that whatever goes wrong in forests and waterways in North America, or in Europe with its disappearing woodlands, or in Asia, or in just about anywhere the Industrial Revolution has taken hold, the damage won't be irreparable. We have to hope that soils won't deteriorate to the point that regrowth of a healthy, diverse plant cover becomes impossible; that forests won't suffer to the extent that they fail to play their part in the ecological cycles that move water or carbon or oxygen from place to place; that the waterways will remain healthy enough, for the most part. Against those hopes lie the consequences of being wrong. It will cost plenty of money to reduce the risk of major environmental damage, and there is a real chance that merely reducing acid deposition will not be sufficient, that acid rain might not be the most harmful of our pollutants. But we won't *know* any of this until the worst, or the best, has happened and we are left to coexist with the results. And so if the soil does go bad, if the forests do disappear, if streams and lakes do die, then we lose. The world, the eastern United States, will have become a much worse place in which to live, one less fertile, one less beautiful, one less hospitable. Place your bet; make a choice.

I don't have a hard time picking sides, not at all. Picture an ordinary rainstorm, an ordinary August morning, a patch of land covered in forest, just as it was two hundred years ago. To this bucolic portrait of the glories of the New England countryside, add layer after layer of a waste product, more each year, and then ask ourselves, "Do we choose to slow the pace or not?" That seems to me like a hell of a nerve.

The Reagan administration's claim that the gaps in science compel it to delay acting on this issue is a blind, a decoy to deflect attention from its simple, political decision. But the language it used to justify its decision reveals a fundamental failure to understand what science can do, ought to do, in our society. The words used to defend that decision echo long after they have served their initial purpose. And this time the echoes sound with something of the cost of scientific revolutions.

This time the cost involves a form of abdication, a surrender of responsibility. Science has become a kind of oracle: if scientists agree, then it must be so; if they are ignorant, we are powerless. A highly sophisticated three-dimensional model is currently being built that will simulate acid storms over the entire eastern United States, a tool that will help identify the most damaging sources of pollution. Rudimentary efforts to model ecosystem-scale responses to stress are under way. There is the entire panoply of more conventional research, from fish censuses to the ancient tactics of raising plants and animals while figuring out ever more ingenious combinations of factors with which to kill them in the controlled environment of the lab. All of this research will help; within a decade or so we will understand the whole issue of air pollution better than we do today. Perhaps our certainty that the tools of the new sciences are so powerful tempts us to wait for the next increment of knowledge in the hope that with that knowledge will come revealed wisdom and the mandate to act.

It is a false hope. A small 'j' judgment day will arrive when perhaps we will know what we ought to have done—when the damage to a forest is irremediable. And if we allow that to happen, then we will be doubly to blame: first, for abandoning our powers of reason, our ability to choose among options with imperfect knowledge, and surrendering ourselves to a faith in experts that the experts themselves disavow; second and more basically, for abandoning the core of our responsibility for the problem in the first place. I drive an old car, a little noisy kid's car with four pipes out the back, spewing out plenty of exhaust. As I headed up I-93 to Hubbard Brook, I was struck by the fact that with each mile I added my increment of nitrogen, some portion of sulfur, and who knows what else to my surroundings. I generated some acid deposition that trip, almost certainly. I drive each day to work, and once there, I sit in an air-conditioned office, and it takes a fair amount of power to keep that running. I play a stereo loudly and often; I go skiing and ride chairlifts; I love looking at tall buildings floodlit into the night. All of us in the day-to-day course of life make our own, individual contributions to the pollution our society as a whole creates. Again, perhaps the scientist's habit of looking at the aggregate obscures this fact; the trick of vision that measures the average acid deposition across the eastern United States and correlates it with sulfur emissions weighted by region and year makes it hard for acid rain to seem a matter of prosaic, personal daily life. But so it is.

In this sense the advances in science actually inhibit our ability to confront issues in which human activity directly affects the functioning of the climate machine. The implication is that we have to choose not just to slap scrubbers onto the worst smokestacks but to alter our ordinary habits—a matter of political decision, not scientific advance. But something is there to be gained from all these high-powered people studying the weather, and it comes back to the question of belief and how we see our world. What the new approaches to climate science require is a belief in the persistence of a web of connections between climate—even the swift and ephemeral ghosts of the weather, like an acid fog in August in New England—and the ecosystem that it supports.

As a city dweller, I am used to the idea of a man-made world; I live surrounded by brickwork and asphalt, with nature bounded strictly by the sidewalks around the park. Acid rain is a product of that world, of the pressures of a growing population and expanding industrial economies. From factory, from automobile exhaust to New England hillside: in this sense what we build extends beyond the limits of the town. With every building and road, we create a cascade of effects that extends far beyond an architect's vision of steel and concrete. The man-made world reaches beyond the city into its—our—surrounding environment and remakes it, under the pressure of all the things we do to keep ourselves in motion, warm or cool, entertained.

And finally, from vision to what we actually see. One measure of the identity we claim for ourselves comes in the monuments we build. Europe in the Middle Ages had cathedrals with which it staked its claim to grace. We have built as solid measures of our greatness skyscrapers and giant factories, and with the factories we have built smokestacks a thousand feet high. In so doing, we leave our traces wherever the wind carries our wastes, and those marks are our monuments as well. Perhaps this is the final reason I choose the side I do: I do not want to define myself as one of those whose greatest monument is a blight.

In August, when the weather clears, the Pemigewasset River is a pleasure to watch, to stand beside and hear as it passes. The forest looks hale, the mountains old and comfortable; it is a place where it is easy to sit and be quiet and be content. Gene Likens takes me out back this August day and spends some time, facing away from his rented condo, orienting me. The bank opposite us is still forested down to the river. Where we stand was bulldozed out last fall to make way for the construction. The builders left one birch tree behind, by design, and it is now neatly framed in the condo's French doors. The tree is still fairly young, but it looks sick; its leaves are scraggly and already some have yellowed. It is simple, says Likens: birches don't live long without the forest canopy around them, and this one, left standing with all the goodwill in the world, is not long for this world. Here is an image, not proof; a symbol, not the thing itself. But I remember the forest across the river, which may or may not be sick, and I remember the rain that fell as I traveled to and from the forest, and I remember a solitary birch that will probably be dead before I see that place again.

CHAPTER 11

When the Desert Grows

IN 1984 the rain did not fall at all, for practical purposes, across the broad band of Africa just south of the Sahara, a region called the Sahel. 1985 and 1986 were better, but not much, not enough. I have one photograph of the area in which I can just make out Tombouctou—Timbuktu as I used to know it—in the background. In the middle ground one person dressed in a dark robe stands along the crest of a sand dune staring back along the ridgeline. In the very front of the picture clumps of thorns lie half-buried in the sand. They look like, and in fact they are, the desert's answer to the grouts with which we armor our coasts to halt the steady advance of the sea. For the sand is moving, and it can kill the land that lies before it.

In the Amazon new roads reach into the jungle. Following the roads come land-hungry settlers and cattle barons. All of them carve their pieces out of the forest, and they come in numbers large enough to mottle whole provinces with their clearings. The land they raze still enjoys heavy rains—it is not a desert by any common understanding of the term—but the cleared land supports fewer species, a smaller weight of plants and animals per unit area; in the hands of the settlers, the land gets worse, grows less. In the Texas panhandle, the ground beneath some counties is running dry. Groundwater free for pumping has been mined away. Fifty years ago the top several inches of the high plains dried up and blew away, as people remember who lived through the black blizzards of the dust bowl.

The drought of the 1980s in Africa has been so stark and so extreme that it almost demands that we regard it as a supernatural intervention, the Act of God theory. That level of disaster, the extraordinary degree of suffering experienced throughout the region for so many years now, must be a trial imposed on its victims. No one would will such a fate on himself, anyone would act to avoid it, and above all, the precipitating cause of the crisis is natural, not the result of human policy. There simply is no rain. Water has always been the gift of the divine, and people don't decide whether or not the rain will fall; they simply enjoy its benefits when it comes or endure the times when it does not.

But the Sahel, which demonstrates this so clearly, demonstrates the reverse as well. Man makes of his climate what he will. Climate, the weather, does not simply happen to human society; responses to climate variation or changes in the way people try to survive within a more or less constant climate determine whether a shift in a climate pattern will become a disaster or simply an ordinary, unexceptional day, or year, or lifetime. If the Sahel provides the extreme case, the easiest to read, it also serves to illustrate how less obvious climate disasters—human disasters—may be made.

In the Sahel, drought is a feature that has appeared throughout history, on every time scale. Contained within the rock strata is evidence that the region was intensely dry 15,000 years ago and stayed that way for at least 2,000 years. It was relatively wet 10,000 years ago, but arid again about 7,500 years ago. Lake levels rose across the Sahel up to perhaps 5,000 years before the present. Since then there has been a slow, general reduction in the amount of water present on the surface of the landscape.

To take it to a time scale that human memory can accommodate, direct documentary evidence indicates climate swings in the Sahel dating back about 500 years. Chronicles preserved from this time tell of relatively rainy periods during the sixteenth and eighteenth centuries, but they also tell of several severe droughts. Drought struck in the early parts of both the nineteenth and twentieth centuries, with wetter intervals in between. Crucially, the 1950s were wet across the Sahel, more or less lush times for crops and grazing and timber—all the natural systems that support the people there. Starting in the mid-1960s, however, the rains became increasingly sparse, and at some point the skein of cloudless days became drought, and drought became famine, and the famine turned the normal modulation of climate into disaster.

Even accounting for the short lives of human beings, the arid lands of the Sahel are so close to the desert that its peoples must have had—did have, the records attest—the knowledge that their climate was subject to this flip from wet to dry and back again. And those societies must, by virtue of their continued presence in the area, have been able to adapt to these conditions in the past. So what makes the current drought different from all others? It's not the climate, surely, for it has remained consistently variable. The disaster—not the drought itself, but the real human tragedy—must therefore have had its origins in human actions. If the climate is behaving normally, the abnormality has to lie in the human response to that ordinary African crisis of no rain.

It is this concept of a human-climate relationship that complicates the science of drought. Most simply, drought, defined as a shortfall in rain, a drop from the "average" precipitation for a region, is purely a manifestation, year to year, decade to decade, up to the millennial time scale, of a climate machine known to vary, probably randomly. The climate system remains within bounds that are broadly set by the amount of solar energy available, interactions between continents and oceans, and so on, but within those broad limits the term *average rainfall* is a myth: the rainfall in virtually every year is above or below the "average," and some seasons, years, or lifetimes will see nothing but drought. This is the simplest definition of drought, one that at least intuitively we all have held. But the evidence of the Sahel forces us to alter that definition of drought and thus, by extension, alter our understanding of the nature of climate and of a human being's ability to mold climate, for good or ill.

The first such reinterpretation of the term is a simple extension of the traditional understanding of drought as referring solely to hydrological drought. Rainfall, to do any good, has to enter the hydrological cycle—it has to penetrate into the soil, enter the water table, appear in wells or rivers and lakes, reach the roots of plants, and then transpire through plant leaves back into the atmosphere. When the ground is bare and compacted, rainfall, even in amounts that would have seemed abundant at one time, can run off the surface and so effectively escape the cycle. Instead of reentering the atmosphere gradually through the regular metabolic action of healthy plant cover, some rain evaporates swiftly, and more escapes the region altogether, flowing swiftly when the rains come, not at all when they don't. The plants that that water might have supported will fail. In this sense of drought, the ultimate result, desiccation, is the result not of a simple shortfall of water but of the operation of a positive feedback loop, in which the impact of human actions can be amplified to deepen the drought. Poor land-use practices can alter the rate of runoff or absorption of rainfall by a given piece of land; if human action alters the hydrology of a piece of land, that in turn lessens both the amount of water that reenters the atmosphere from that turf and, in theory, ultimately affects the amount of rainfall downwind. In this scenario, drought becomes not simply a shortage of rain but a restructuring of the entire climate system, which is now molded by the purposeful agency of the dwellers on the land rather than by the random fluctuations of the winds and ocean currents.

In the Sahel itself, the particular acts that have deepened the disaster will make for a sad tale of error when all the facts are in. The World Bank, in an in-house report on desertification in the Sahel, somewhat disingenuously argues that overgrazing, deforestation, urbanization, the extension of rain-fed farms into the bush, the extension of roads, improvement of medical and veterinary services (leading to human and animal overpopulation), the breakdown of the traditional family, creation of centralized governments, and the growth of the money economy have all contributed to the abuse of the landscape. When drought struck, in other words, the region had no ecological safety net—no spare pasturage or healthy stands of forest—to help the local population overcome bad years. I say the report is disingenuous because the World Bank itself had a hand in at least some of these developments, which it, understandably enough, does not make much of in this particular report. But leaving aside blame for a moment, this list of things done wrong in the Sahel suggests that only now are people beginning to realize how the climate, in human terms, actually operates. The climate is defined by a relationship between the weather, rain or no rain, and what use people have tried to make of the natural world. The inhabitants of the Sahel were not able to exist independent of their climate, able to impose their will—by increasing their herds or growing crops where grasses had grown before—without eliciting some response from the land itself. Its response was to become less resilient when the ordinary, recurrent drought struck.

We are a long way from the Sahel, but that blasted land is the canary in our coal mine. It is so close to the edge, ecologically speaking, that its inhabitants have become the Greenland Norse of our generation, victims of the disaster possible when people try to impose a way of life upon a climate system that will not sustain it. In other regions people may not die quite so readily as a result of a change in the weather, but the history of the Sahel should tell the rest of us that the interaction between climate and society can definitely be changed for the worst.

Baldly, whenever I stop by the road to buy a hamburger, I wreak my own small increment of havoc all over this hemisphere—I make the land poorer and less able to stand up to the weather. Consider an ordinary fast-food burger: To obtain beef, one needs cattle; cattle require pastures; billions of burgers' worth of cattle, or however many may now have passed over the transom, demand an enormous amount of pastureland. The United States produces plenty of beef, but for sufficient quantities of cheap meat several of the fast-food chains have turned to huge Latin American cattle operations, which can run what seems like unlimited numbers of cattle on what seems like a limitless supply of land within the more or less undeveloped rain forests that cover the Amazon basin.

Loaded words, *seems* and *limitless*. South America has vast forests but not uncounted acres, and each burger cuts its part out of the whole. Two researchers, Christopher Uhl of The Pennsylvania State University and Geoffrey Parker of the New York Botanical Garden's Institute of Ecosystem Studies, came up with a formula for calculating the direct exchange rate between hamburgers and the land. They began with the weight of all the plants on a single average hectare of rain forest—about 800,000 kilograms, dripping wet. A hectare of pasture cleared out from the rain forest will produce about 50 kilograms of weight gain for the cattle run on that piece of land. Such pastures generally last about eight years, after which they cannot sustain even the grass cover needed to support grazing, so over the life of the pasture that land can be expected to produce over 400 kilograms' worth of cow. Accounting for skin, bone, and other such detritus, that leaves about 200 kilograms of beef as the total output of that unit of land.

These are crude numbers—more calculations of the general order of the problem than anything precise. But from here on, it is simple arithmetic: 200 kilograms of beef translates into 1,600 quarter-pound burgers: dividing the 10,000 square meters that make up one hectare by the 1,600 burgers it produces shows that it takes 6.25 square meters of pasture to produce a single hamburger. Uhl and Parker then speculated just what that plot of forestland might otherwise hold: one 60-foot-tall tree with perhaps fifty saplings and seedlings, insects and animals, microorganisms, mosses, and fungi. "All told," write the two scientists, "there are millions of individuals and thousands of species" associated with the land it takes to produce one hamburger. In a straight mass for mass calculation, it takes—or I consume—a half ton of forest with each quarter of a pound of beef.

As a climate phenomenon, rather than simply as a land-use issue, rain forest-for-pasture trade-offs are significant because on a large enough scale logging out the forest fundamentally depletes the ecological richness of the land. Rain forest soils are, paradoxically, quite thin, for all the diversity of life they support. Although the humidity of the rain forest encourages the rapid decay of any dead plant life that could compost on the forest floor, the rain forest itself is so vigorous that it locks up most of the available nutrients in living plants. Without the protection of the full forest canopy and the support of root structures, though, the soils deplete rapidly, eroding away with the rains and swiftly losing what nutrients remain after the destruction of the forest cover. This process accounts for the short eight-year life span for pastureland cut out of the rain forest that Parker and Uhl used as a basis for their calculations. Given enough time, a secondary forest will regrow, usually, but it will be far less rich and vigorous than the original ecological community.

Even more directly, the plant cover helps determine the weather. Some areas of the Amazon basin receive 200 inches of rainfall each year. Half of that water, if not more, is falling on the forest for at least the second time: the escape of water vapor from the tree canopy supplies the hydrological cycle of the entire region. Cut down enough trees and you could affect rainfall downwind of the deforested area—leading once again to the possibility of hydrological drought. Although a correlation hasn't been firmly established yet, some Brazilian observers have argued that deforestation in Amazonia is contributing to a loss of rainfall in the agricultural regions of the country well beyond the forest margin.

And yet no one can actually tie a drop in rainfall here to a cattle ranch carved out of an Amazonian province there. The slash-and-burn agriculture practiced by indigenous people has logged over huge areas of forestland for hundreds of years; what's happening now may be causing damage, but plenty of people argue that it is fundamentally no different from the long-established—and thus presumably sustainable—pattern of human behavior. It is almost impossible with conventional methods even to state how much land is being deforested each year, much less attribute any consequences to the aggregate loss of forest cover. Perhaps regions are drying out because of chance fluctuations of wind patterns; perhaps they are not dessicating at all but are suffering a momentary dry spell that might be no more than a blip of drought on the climate record between periods of lush rainfall.

Thus, even if we suspect that cutting down enormous tracts of rain forest is a bad idea, such conventional tools as rain gauges and forest censuses cannot even begin to measure how bad an idea it is or what the long-term consequences may be. What is needed is something simple, some fundamental change in vision, in the way we are able to look at our world and pick out one event, a drought, for example, from the background noise of a complicated system daily going about its business. Such new "eyes" exist, fortunately, and they are wonderful in the sense that their capabilities, the power with which they can absorb and distinguish information and events, are a source of awe. I am amazed that the simple act of looking, when perspective and the power of our vision change, can transform what we know; I'll leap ahead of myself to say that I hope that such an extension of our sight may change the way we act as well.

Begin with the images these new tools record. In one of the first applications of the new tools, a series of pictures was generated, as famous in its own circle of onlookers as the *Mona Lisa* might be, or for a closer analogy, as Picasso's painting of the attack on Guernica is to a general audience. There are three images, mottled red. On a rough diagonal, running upward from right to left, is a fine, slightly crooked line, standing out starkly white. Running off that main line, short crosshatches score the image. In the oldest of the images, from 1976, the white lines are whisker-thin, almost negligible against the red background; by 1978, the white lines are deeper, wider; and by 1981, the lines have become patches of white that have begun to merge.

The images are recordings made by Landsat satellites, which sense radiation in infrared wavelengths—heat. The red areas mark forest cover, where the trees absorb light and reflect near-infrared radiation back out to space (and to the monitoring satellites). Bare ground reflects more visible light and less infrared radiation than does a healthy forest, and cleared patches show up as white areas on a Landsat image.

The pictures chart a square 185 kilometers a side in the Brazilian state of Rondônia, along the southwestern end of the Amazon basin. The major line across the image is the Balem-Brasilia Highway. Built in 1960, it opened to year-round traffic in 1967 and World Bank funds paid for its paving in 1974. The crosshatches are the access roads built to allow settlers to carve homesteads out of the forest. Before the highway was paved, about 1,200 square kilometers of the total 243,000-square kilometer area of Rondônia had been deforested. By 1978, four times that area had gone under the blade, and by 1980, the amount deforested had almost doubled again. Similar measurements, using data from Landsat and the National Oceanic and Atmospheric Administration (NOAA) weather satellites, showed the same trend in other states in the region: 8,000 square kilometers gone in 1975 in Pará, 34,000 square kilometers cut by 1980. In Mato Grosso, 10,000 square kilometers deforested in the beginning of the period, 53,000 cut down after five years.

These figures should be placed in perspective. The entire area legally designated as part of the Amazon consists of over 5 million square kilometers. A few thousand square kilometers cut here or there would hardly make a dent, it would seem, and by any measure the great bulk of the Amazon is, even today, largely wilderness. Images made by the side-looking radar on the space shuttle form a giant scroll, much like the great traveling pictures of the nineteenth century that would unroll to reveal life along the entire Mississippi, say. These pictures are black-and-white, instead of the false colors of Landsat images, but they track clearly enough a narrow swath all the way across the Amazon basin. For mile after mile the view is the same—a mottled gray showing a healthy forest. While bare patches, indicating development like that taking place in Rondônia, show up, the impression is of a huge and healthy forest, with only gnat bites taken out.

But the snapshot, the static image deceives: the forest may look intact, but the trends imply that it is not, or that it will not be so for long. Philip Fearnside, an American scientist working at the National Institute for the Study of the Amazon in Brazil, told a symposium of ecologists at Woods Hole, Massachusetts, in 1986 that

> the shape of the growth curve of the deforested area is crucial—the most dangerous tendency is for the areas to increase exponentially [that is to say, the area may be doubling over every survey period]. . . . The difficulty of intuitively understanding exponential change is great, even for we who live daily with a phenomenon such as inflation. Thus, for many people it seems impossible that the relatively small deforested area of the Amazon today could increase within a few years to the point of encompassing the whole region. This is precisely what would occur if deforestation were to increase in an uninterrupted exponential fashion.

This is the power of seeing, of vision extended, of the evidence of images collected impersonally year after year. At ground level the forest always begins at the edge of the clearing, and no matter how much one clears, the forest never ends. There is no way from this perspective to perceive, understand, or, to use a word unfamiliar to science, to *feel* the impact of each blow of the axe, each swipe of a bulldozer. But look down from above, calibrate the instruments, and measure those relatively tiny increments of heat absorbed or reflected by forest and bare ground, and then the danger becomes clear. The satellites have joined the computer as the central tools of climate science by making it possible to see, for the first time, not just a single pasture or a ranch project but the forest entire. It has long been recognized that clearing a single patch of rain forest does that piece of land no good, at least for a while; until we had gained the technology to peer from space, we had no way to judge whether the forest as a whole was at risk.

Most important for the advance of climate science, satellites, properly used, do not simply gather more information of a sort we already have; in the right hands, the change in scale from local to global exposes processes that were previously unsuspected. At NASA's Goddard Space Flight Center in Maryland, Compton Tucker is a scientist driven by what the satellites offer him. His ultimate aim, a Napoleonic one, is to build a complete picture of the day-by-day functioning of the landforms on every continent on earth. For now, he has undertaken a slightly more modest program, and has managed to create the means to measure the primary productivity of a piece of landscape: that is, he can, by using satellite data already available, image a region of a single square kilometer and determine how vigorously the plants on that site are growing. He can, for example, measure the aggregate amount of photosynthesis—carbon fixed, oxygen released—in the plant cover on a wheat farm in Kansas or on a patch of land in Rondônia.

Tucker's findings are based on an elegant and fairly simple technique. NOAA's weather satellites have sensors that recognize four different signals: one sensor detects visible light, one recognizes near-infrared radiation, and two more sensors recognize other wavelengths in the infrared band of the electromagnetic spectrum. The pigments that turn plants green are contained in chlorophyll; the greater the amount of chlorophyll in the plants, the higher the rate of photosynthesis. Green pigment shows up strongly in visible light, so green plants look really green, and we, and the weather satellites can see it. However, green absorbs rather than reflects the near-infrared radiation. So a piece of landscape with plants vigorously growing on it would appear very bright in an image made by the visible-light sensor, and very dark in one made by the near-infrared one.

By comparing the visible and near-infrared images, Tucker is able to eliminate the possibility that one or the other sensor in the satellite has drifted off its original settings, to match up and account for the presence of clouds, and to come up with an accurate picture of how well plants are growing at any given spot at any given time. To Tucker, the patches of deforestation spreading like mold across Amazonia are visible not as places where trees have been removed but as landscapes whose productivity, judged by the most basic measurement, has declined. He has performed the same analysis of images made of the Sahel, weighing the severity of the 1984 drought by noting how far to the south the band of land that supported no photosynthesis moved in that year. From this perspective, drought is not regarded as a shortage in rainfall or a drop in the water table but "as a deficiency in photosynthesis."

By this reckoning a significant climate event occurs when, for whatever reason, the biosphere—the ecosystem, the plants—changes. With this definition, drought and hence climate take another step away from the simple calculations of weather. Drought becomes not some straightfoward physical measure (ten inches of rain fell last year, not the normal twenty-two) but instead reveals itself as the end result of interactions between the weather, the landscape, and people. So while Tucker identified the drought in Africa by the disappearance of plants, an event clearly precipitated, proximately, by the lack of rainfall, he can also detect what one could call a drought in the Amazon, where the cattle ranches and the spreading settlements in the rain forest appear in his images as a mottling of light and dark, as land no longer supports the richness of life that it once could.

Knowing this, however, is not the same thing as being able to predict drought. At best, what Tucker's data permits is a kind of commonsense approach. It shows that if we keep doing to the Amazon what we have been doing, then the forest will decline, less photosynthesis will take place. Given all that, it is at least plausible to conclude that rainfall patterns will shift and by some definitions (such as Tucker's own definition) a drought will occur. The prediction is born of the satellite images and of the rising curve of forest destruction those images reveal.

That is the underlying message of drought. Prediction, in fact, is not the most important issue; after all, we know that some years will be better than others. The goal is to retain sufficient flexibility to bend with the wind—and to preserve sufficient ecological richness to allow the land to snap back to optimum conditions when the difficult years end. What has stood in the way of achieving that goal is the conflict between a common gain from some resource or system and a private profit that can only be met by abusing the larger interest. Such a clash is as old as human history, but now that conflict, driven by unprecedented population pressure and the extraordinary technology now available, threatens to engulf huge tracts of land and the lives of those who live upon them. Recognition, if not understanding, comes from seeing.

Here is one more image: The date is 14 April 1935: Black Sunday. On that day a huge cloud of dust blew across the midwestern plains, blew across Kansas and into Texas and New Mexico. Historians have turned up macabre details from that day. Donald Worster writes in *Dust Bowl* that in Guymon, Oklahoma, the Reverend Rolley Wells held a rain service at the Methodist church in town and finished just early enough for his congregation to make it home before the dust storm hit.

And there are the photographs. The one that most frightens me is of a clump of houses and outbuildings somewhere in western Kansas. These are houses of poverty—little one-story wooden structures that even in the old photo seem to need some paint, two- or three-room places with a couple of small windows cut into each wall. They look terribly vulnerable, and looming over them like the wrath of God, just about to engulf them, is this enormous, billowing black cloud.

The statistics on the black blizzards of the thirties are terrifying enough. From the five giant storms in the spring of 1935, the dust fallout was estimated at over four and a half tons per acre. But it is the picture, and not the number, that resonates. Drought to Americans, at least to Americans of a certain age, means the dust bowl, and the dust bowl is that place in the imagination in which black storms sweep down the plains to engulf the little houses and all the unfortunate people in their way.

The problem, as in the Sahel, was lack of rain. One standard measure of drought compares the actual rainfall during a period with the expected average rainfall. Using that measure, by the end of the thirties, according to Worster, some areas of the Great Plains lacked more than thirty inches of precipitation—almost two years' worth of rainfall. Extraordinary, and explanation enough for hardship. It is difficult for grain, beast, or man to live if there is no water to be had. But like the Sahel, the dust bowl region is one in which what we call drought, an abnormal phenomenon, is an ordinary, if irregular, feature of the climate system. The dominant climate forces are pretty well mapped out for the area. It gets its weather from three major circulating systems in the atmosphere: one over the Gulf of Mexico, one bearing polar air, and one marked by the prevailing westerlies, which drop their snow on the Rockies and hit the leeward side bone dry. In years when the westerlies gain strength enough to push wetter air out of the way, very little rain falls on the bowl.

In the now familiar fashion, knowing how the weather works doesn't directly help in predicting when it will work, in which way, and for how long. Certainly the thirties were unprecedented in the history of the agricultural use of the Great Plains. In the late 1880s the prairie grasses went under the plow; between the sod busting and the black blizzards four decades later, agriculture boomed. The farmers that suffered first and most in the dust had no experience with the disastrous consequences of a combination of shortfalls in rain and cleared soil, and thus they had no common pool of memory or expertise that could prepare them for the black blizzards.

No memory, perhaps, but the farmers collectively bear some responsibility for the advent of the dust storms. Plains grasses do not blow a thousand miles to rain on Kansas City; dirt does. The dust bowl, if nothing else, provides an object reminder of the risks of wholesale changes in an ecosystem. The drought of the thirties was a natural event, but the dust storms that followed in its train are examples of weather made by human beings, who with their tractors turned the plains into a kind of dirt mine, ready for the winds to ship the product out.

And yet that was long ago and far away—we aren't in Kansas anymore, not that Kansas. In the broadest, most oversimplified terms, the causes of the dust bowl are easy enough to pinpoint. Worster argues that the creation of commercial agriculture on the Great Plains led to several dangerous practices. First and foremost was the simple act of sod busting, which increased dramatically with the introduction of the tractor in the first decades of the twentieth century. Not only did tractors break up the sod with amazing efficiency, they made it possible to plow horizon to horizon, running up and down in straight lines. Unlike the Amazon, where deforestation occurs in patches, precariously hacked out wherever a road gives access through the jungle, the endless expanse of the plains offered almost no barriers to a determined tractor driver. It became possible, in effect, to remove all of the grass over huge stretches of ground.

We have learned at least a little from our profligacy then. Severe drought struck the region in the 1950s, and drought comparable to that of the 1930s returned in the mid-1970s and again in the 1980s. That most recent drought did enormous economic damage, but the black blizzards did not blow again, and grain surplus, not the collapse of the harvest, became the byword for the region in the late 1970s and 1980s. People had by that point planted windbreaks, and contour plowing—plowing along the lines of the terrain, not against them—had become increasingly widespread. Above all, farmers had begun to irrigate. Although relatively little rain falls on the dust bowl, much of the area does lie over the Ogallala aquifer, an enormous store of so-called fossil water. Water enters the aquifer in a small area that catches a good deal of the Rocky Mountains snowmelt; over more than hundreds of thousands of years, that water has seeped beneath a large portion of the high plains, from Nebraska to Texas.

Irrigation with Ogallala water began in earnest after the Second World War. From 3.5 million acres under irrigation in 1950, the total reached 15 million, or a little under half the area of what had been the dust bowl, by 1980. Running irrigation wells is an obvious way to deal with drought—if the rains don't fall, water in abundance still reaches the fields. The catch, of course, is that wells eventually run dry. In the southern part of the Ogallala aquifer's range, that has already begun to happen. In Texas and New Mexico supplies from the aquifer are beginning to be depleted, as they are in scattered areas throughout the high plains.

Jonathan Taylor, then with NCAR, studies how people respond to drought. He went down to Swisher County, Texas, to see how local farmers were preparing themselves for the next dust bowl. The most striking fact to emerge from his time there is that plenty of people have a sense of the danger, but their collective actions place them smack in the path of it. Even more troubling, each individual response to the threat makes sense for that individual, but add up all the decisions and everyone involved loses.

Taylor remembers from his visit two anecdotes in particular.

There was one farmer he wanted to interview, a man active in the local community and known for running a tight operation. The man didn't open much to Taylor—wasn't home, didn't offer to make time to talk, made himself simply unavailable—so Taylor drove out one day and just met the farmer on his spread. The farmer was in the process of running a new well through an old pipe system; since pulling a well is at best a two-man job, Taylor just took off his coat and helped out. With a favor to return, the man finally consented to talk to the geographer. The farmer told him that he was trying to build an insurance policy against drought. The well the two of them had just placed was a two-inch pipe— essentially a household well—run within the ten-inch bore of the old irrigation system. The idea was to convert to dry-land farming, which meant relying on the annual rainfall, but to retain a small backup system that could drip water onto the land only in times when the rainfall ran short.

Taylor regarded this man as one of the smartest and most efficient of the farmers he met in the course of a survey that took him through five states. Another man with a farm near this first farmer's land also impressed Taylor as a clearheaded planner. He faced virtually the same circumstances as the first man: he farmed irrigated land over an aquifer that was becoming progressively more expensive to draw water from, and the end of affordable water was clearly in sight. His response was to irrigate more, to pump up as much water as he could use each year, and to raise as much of a crop as he could every growing season. His reasoning was that he would soon be forced to switch to dry-land farming and that in order to maintain himself and his family at the standard his irrigated farm afforded them, he would need a much larger dry-land spread. So for the present he has decided to make as much money as possible, expand as much as he can, so that when the aquifer in that area really and truly does sink past the level at which water can reasonably be mined out, he'll own land enough to sustain himself.

The tragedy here is twofold. First, and most obviously, the strategies these two near neighbors are following make perfect sense for the individuals trying to follow them, yet each is absolutely incompatible. As long as the second farmer sees a way out of the bind that his climate and his previous farming techniques have landed him in, he'll go on pumping. The first farmer could perhaps make a go of it, but only if his little two-inch wells have something to suck up; given his neighbor's strategy, they won't. That will leave him completely vulnerable to the kind of disaster that once already this century blew the top layer of Kansas across the Mississippi to points east. The formal definition of this kind of drought is agricultural drought; the weather patterns remain the same, but the human use of the land has become mismatched to the climate of the region, even after the application of technology has modified the water equation.

That is the second tragedy: As water sources dry up, the odds that the black blizzards will return rise. Texas Tech University's Harold Dregne, one of America's leading drought specialists, states unequivocally that the black blizzards will return, perhaps with the next great drought, or maybe with the next drought or the one after that. Whenever they do reappear, it will be the fault of human beings, a climate disaster that is man-made.

In the Sahel, when drought occurs and the human societies there fail to accommodate themselves to their climate, people starve. If the dust storms retake the plains, most of us in the United States will probably survive comfortably. No one is likely to die of hunger as a direct result of a drought in this country in the near future. We're rich enough—we can buy beef from Latin America; we can grow plenty of crops in the parts of our broad land that such droughts would miss; and we could find the cash to import grain for a time should we face the unlikely fate of an absolute shortfall of food. But the underlying fact will remain: human misuse of the land will have turned regular features of regional climates into human catastrophes.

Different tragedies have different emotional colors to them. It is tragic when a rock shaken loose by an earthquake causes the death of a child, for example. It is enraging if a child should die in a quake because the concrete in his home was more sand than cement. But the kind of region-wide, causally complex disaster, such as a major drought, inspires less rage, less of an obvious course of action (jail the contractor) than do more individual misfortunes. Climate disasters still appear to be acts of God, and the scale of desertification and drought is too large to apportion individual blame. I cannot say that Farmer X is responsible for the dust storms to come or that one particular ranching operation is going to be the one that finally tips the balance in the Amazon. But the truth of the matter is that these and a hundred other examples of landscape impoverishment are avoidable. They don't have to happen.

Specific actions could be undertaken, or undone, to lessen the potential dangers. Clearly the kinds of development practices the World Bank itself criticized in its report on the Sahel should be avoided. Don't overpopulate the carrying capacity of an era, with beasts or men; don't carve roads into regions that cannot support the traffic those roads suddenly become able to bear; don't pump our wells dry to grow yet more of a crop surplus. Environmental groups have begun negotiating with the major development banks to get them to at least consider the ecological consequences of major projects before going ahead with them. In each area slated for development, particular choices can be made that will differ, depending on the character of the place. Political decisions must be remade or unmade: the policy of the Brazilian government to encourage settlement in the Amazon is an obvious target; the U.S. government could conceivably ban the import of Brazilian beef as a disincentive. That would cost us, as well as the Brazilians, but all these issues point out that a different kind of accounting will have to come into play. It isn't good enough simply to tot up how much money can be made from a year's wheat crop; the cost of long-term damage to the land has to be considered, and that consideration may mean—will mean—that some actions that are profitable now should be forbidden.

I don't have a lot of confidence that such hard choices will be made. One overarching problem links the Sahel with the Amazon, with the high plains, with all the regions and local problems from California to Australia to the "green deserts" of India, and on elsewhere, that I haven't talked about here: the persistence of the belief that human societies exist outside of the natural world and can impose on that world the shape that most suits them at any given time. That belief anchors the creed that any person may make what use he will of the resources the natural world offers him, whatever the long-term consequences (or even short-term consequences) to himself, his neighbors, or to people thousands of miles away.

Science can give guidance here, but no solution. With new machines and the work of people like Tucker and others, it has become increasingly easy to identify the problems, to notice when in fact our drive to expand has reached a scale large enough to adversely affect whole regions. But the role of climate science is circumscribed by the nature of the climate event it investigates. Essentially, the tools of climate science can record clearly the status of a living system—when plants are thriving, when they suffer. In addition, they make it possible either to identify the degree of variability in a climate region, as with the climate history of the Sahel, or to model and thus to provide at least a theoretical understanding of the long-term climate consequences of the feedback loops between changing plant cover and an alteration in the weather, as students of the Amazon are trying now to do for Brazil. Drought prediction or drought prevention, though, is another matter. If drought were simply insufficient rainfall, then the prediction problem would be the same as that for any weather forecast. Once the matter becomes not just water, but water and what you do with it, then what becomes important is not the weather but the human beings who live with, or rail against, the sky.

But even recognizing the phenomenon of drought when it occurs, in whatever form it occurs, is crucial. The ability to see what is happening to our environment and to understand it is emerging, in fact, just in time. The pace of change introduced by technological progress is such that if our attention lapses for even one generation, a smidgen of time on any geological scale, transformations in the land that are truly dangerous could catch us unawares. The satellites up there and the computers down here monitoring the metabolism of the earth now make it possible to gauge how and where the greatest changes in the land are taking place.

That knowledge is only the spur, of course, and not a conclusion in itself. The evidence to date seems damning enough to dictate policy, but the decisions that might be made have as much to do with a moral sense as they do with any specific scientific conclusion. We need to recognize that certain uses of the land upon which we live are simply wrong.

Frederick Manning drew from his experiences in the First World War this belief: "War is waged by men; not by beasts, or by gods. It is a peculiarly human activity. To call it a crime against mankind is to miss at least half its significance; it is also the punishment of a crime." The horrors of starvation are certainly crimes against mankind; the poverty that impels people to leave cities in Brazil to carve out pieces of the Amazon is a crime; the dreadful experience of living in the dust of the thirties was one that nobody would wish on another. And yet while the individual victims may be blameless, it is not the climate, nor the intransigence of a piece of land that will not grow what we want it to grow, that commits the crime. Our actions form a direct analogy to Manning's experience of war: we make our world a harsher place in which to live, so we the victims must be seen not just as victims but as those doing penance for the damage done to the earth.

Natural systems are enormously resilient: plants adapt to and modify climate, creatures live off and manipulate the communities of plants, trees recycle water that falls on other trees, great landscapes form across all the time scales that affect the lives of single organisms, ecosystems, and geological formations. Human beings have been on the scene, on the various scenes, for a terribly short time: large-scale farming on the high plains has only a hundred-year history; most of the Amazon has never seen major settlement; and more people by far live in the Sahel today than ever before, even in that ancient region of human habitation. The tools available to us now give us the potential to understand what it is that this human presence means; that understanding could in turn give us the chance to make something other than the worst of choices. I am an optimist in that I believe that human beings do not always perversely try to make life worse for themselves. But the margin for error is not unlimited. The English poet Philip Larkin once received a commission from the British Department of the Environment to write a poem that appeared in an official pamphlet called "How Do You Want to Live?" Larkin called his poem "Going, Going," and in it he included these lines:

It seems, just now,
To be happening so very fast;
Despite all the land left free
For the first time I feel somehow
That it isn't going to last.

CHAPTER **12**

A Warm and Foreign Place

DOWN BY THE ELBOW OF Cape Cod, facing west into Buzzards Bay, is a small piece of land that appears to be absolutely insignificant. It is perhaps three-quarters of a mile long by a third of a mile deep. It fronts on dunes and a little outlet that flushes water from the shore into the bay and allows the tides back in again. It is marked off at the rear, on the landward edge, by an old railroad cutting. The regular trains have long since stopped running down this way, but the rampart and the tracks and the roadbed stones remain. There is no ready access to this piece of property—one can't just drive up to it or onto it—but then, there is no compellingly obvious reason why anyone would want to go there either.

It looks like what it is: a low patch of coastal wetland, a marsh—now, in October, mostly brown and dull. At first glance, when one scrambles down from the railroad tracks, it appears to be almost level, covered with spiky grasses, broken only by the occasional hummock and the line of a creek. John Teal is a biologist at the Woods Hole Oceanographic Institution and, as much as any person living, knows the salt marshes; especially, he knows this one, the Sippewisset salt marsh, this insignificant piece of land, this archetype. To Teal's eyes, if not to mine at first glance, the salt marsh is a place of enormous variety and richness; he has been studying this one for the last twenty-five years, because a salt marsh makes a neat frame of reference in which to see how ecosystems change when their environment alters. We are here today to see what will be lost soon, for the Sippewisset salt marsh is bound to disappear because of a climate change caused by human action. It won't be the most important loss, of course, these few acres of grass and peat and moving water, but the marsh's fate provides a template easy to perceive, on which to build an understanding of the greater transformations to come.

Most important, what will happen to the salt marsh, to us, to the world, is something qualitatively new. Climates change on every scale of time and distance. Human beings have bent to change or bent the change to their own ends from the dawn of human history. England used to be forested from Cornwall to the River Dee, then sported vineyards for a while, today grows grain and raises fine young lambs in the fields between its towns. Grasses covered the high plains a century ago, while wheat, at least for the time being, does now. In contrast, modern climate phenomena—acid rain, air pollution, desertification—are happening at a speed and extent that are directly related to the rate at which human beings have bred and spread and increased the amount we consume. And yet for all our numbers and our appetites, we live in particular places, and our actions reverberate first at home. Acid rain falls downwind of where we burn our oil and coal—first on us, then on our neighbors, and so on. One patch of cleared forest loses its topsoil, and then the next patch as it in turn is cut. Huge areas of land may suffer as many peoples make the same mistakes in different parts of the world, but nothing in each of these processes inevitably affects the entire world.

And so it has always been. The Fertile Crescent along the Tigris and Euphrates used to hold such rich farmland that it cradled the first empires and the most literate of the ancient civilizations, the Sumerians. The Sumerians failed to drain their irrigated lands properly, and to this day their fields grow stones. But their neighbors to the north, the Babylonians, prospered for much longer, and Babylon has a stronger hold by far on our memory than do any of the great days of Sumer. But sometime in the last few years—since the Second World War, at any rate—the nature of the relationship between human beings and their climate changed fundamentally. We now can change not only each area in which we happen to settle, but the world as a whole.

We do so by attacking the climate system at the point at which effects can most swiftly be propagated around the globe—in the atmosphere. What is happening is that the by-products of our economic life are altering the chemical composition of the atmosphere by increasing the amount of carbon dioxide and a number of other rarer compounds. These compounds together alter the radiative properties of the atmosphere. The net result is that the lower atmosphere and the surface of the earth are going to retain more heat; within a short time, one generation or two, the entire planet will be significantly warmer.

To which one of the most common reactions is "so what?" or even "fantastic." After all, the warm years of the tenth and eleventh centuries are remembered as the Medieval Optimum, and the prospect of shorter winters, lazier summers, and, more practically, longer growing seasons in the upper latitudes is one that would seem eminently desirable. And in any event the scale of the predicted warming is from about 3°F to 9°F as a global average. At first glance, that is hardly intimidating: plenty of places flourish where the temperature shift from day to night is 30°F or more; and season-to-season averages in the temperate zones, for example, can vary more each year than any increment this supposedly menacing greenhouse could produce.

But the issue again is one of perspective. We simply don't have it. Human beings have never in their recorded or remembered history experienced anything like the scale of the change that our ways of life are now imposing on the global climate system. The transition from the last ice age to the current, largely ice-free world in which human society has thrived hinged on a temperature rise of the same, apparently modest scale—several degrees Fahrenheit, no more. One crucial difference is the speed involved. The shift from ice age to interglacial took a few thousand years; this new, human-driven warming will reach the predicted strength, if the climate theorists have it right, in our lifetime perhaps, during the lives of our children, certainly.

The greenhouse effect is a touchstone, an event so extraordinary that it shapes all the rest of our existence, but the problem is that it is typically an event that happens out of sight. As a part of daily life, the greenhouse effect poses probably the hardest test of the imagination, requiring you and me to leap from the local, our rooted life in a given place, to the grand global abstractions of "the atmosphere" or "the climate," and then back again to each of the particular patches in which we live.

But I don't live in the world; I live in Boston. I don't change the world; I drive my car five miles to my office, and at night five miles home again, burning about a third of a gallon of unleaded high-test on the round trip. The greenhouse effect, this global transformation, does not fit the ordinary categories of experience, and so unless it can be trimmed to size it will remain remote, an abstraction, until it slams us in the face.

Cautionary tales—premonitory courses of events—are already taking place in pockets of everyday life, but first we must look to the core of knowledge that creates a larger context for each specific story. Essentially, what we already know is how the composition of the atmosphere is changing and what the first likely result of that change will be. The research effort has focused on tracing the effects of growing levels of five compounds. Carbon dioxide is, of course, the classic greenhouse gas, with the cycle of carbon dioxide production (from volcanic eruptions, plant growth, and the like) and consumption (on land and in the oceans) implicated in warmings ranging from the earliest days of the earth up to the natural cycle of ice ages and interglacials. To this natural cycle has been added the smoke from modern fires. Woodsmoke contains carbon dioxide; the burning of felled tropical forests is in fact only one of the novel ways twentieth-century society has come to alter atmospheric chemistry. But the consumption of oil and coal since the Industrial Revolution has caused most of the measurable change in the carbon cycle and the buildup of carbon dioxide in the atmosphere.

Carbon dioxide is important as a greenhouse gas because its specific radiative properties allow it to fill a particular niche, regulating the energy flow from sun to earth and back out to space. Water vapor is far more abundant in the atmosphere than carbon dioxide, and it also possesses the ability to absorb infrared radiation escaping from the earth toward space; however, in one part of the infrared band of the electromagnetic spectrum—from 7,000 to 13,000 nanometers—water vapor absorbs that radiation only weakly. Because the earth's energy budget has to balance (that is, all the energy the earth receives from the sun has to be matched, ultimately, by energy radiated out), most of the energy that escapes from the earth leaks out as heat radiating on those wavelengths. The 7,000- to 13,000-nanometer region of the spectrum is known as the atmospheric infrared window. Unlike water vapor, carbon dioxide does absorb infrared radiation within this window; hence, a rise in carbon dioxide levels means that more energy remains in the lower atmosphere, increasing the earth's surface temperature. (The energy budget stays even, however; as its surface warms, the earth radiates the same amount of infrared energy out, thus balancing the heat escaping out to space with the light coming in from the sun.)

In 1957 a young graduate student named Charles Keeling went to a site 11,000 feet up the flanks of Mauna Loa in Hawaii to set up instruments that could detect the absorption of infrared radiation in the atmosphere. That data in turn provided an extremely precise measure of the carbon dioxide content of the air under scrutiny. What Keeling has found in his readings over the three decades since is that carbon dioxide levels have been steadily rising, year to year, and that the curve upward takes a zigzag path: each year the amount of CO_2 peaks in early summer and drops to its low in early winter. Currently there are about 350 parts per million of carbon dioxide in the atmosphere, up from about 315 when Keeling began his measurements. To appreciate just how dramatic a rise that is, consider this: using several indirect measures, a number of researchers have fixed the level of carbon dioxide in the atmosphere around the year 1800 at between 260 and 285 parts per million.

Other gases, all much rarer in the atmosphere than carbon dioxide, also absorb infrared wavelengths within the atmospheric window. In the late 1970s, though, several researchers began to question whether any such trace gases might actually be accumulating swiftly enough to augment the impact of the carbon dioxide greenhouse effect itself. By the early 1980s they had found four: nitrogen doxide, methane, and two from a class of chemicals collectively known as the chlorofluorocarbons. Oxides of nitrogen are another by-product of burning fuels, as is methane. (Methane is also produced in the bellies of ruminating animals, at the floor of rice paddies, and by the bacteria that enable termites to digest wood, all of which makes identifying specific causes of a rise in methane a little difficult. However, because the gross increase of methane in the atmosphere seems to run in sync with human population growth, it is possible to hazard a rough guess as to methane concentration trends, if not to the precise accounting of which source emits how many molecules of the gas.) The chloro-fluorocarbons, especially the two called CFC 11 and CFC 12, are man-made; they are the compounds used as the working fluids for refrigeration, aerosol cans, in parts of the production process for certain plastics, and in some other applications. While these chemicals are substantially outnumbered in the atmosphere by carbon dioxide, molecule for molecule they absorb more radiation, and between them, they are projected to account for as much potential warming as does carbon dioxide by itself.

Given a basic understanding of the actual radiative properties of the five gases and estimates of rates of emissions of each, climate modelers have been able to attempt a prediction about the time and size of the warming. In model experiments, the standard test has been to take a climate simulation that has performed well in re-creating current climate events and then in the model atmosphere to double the amount of CO_2 from the preindustrial level of 280 or so parts per million to 560 parts per million. Groups of modelers at NCAR, at Princeton's Geophysical Fluid Dynamics Laboratory, at NASA's Goddard Institute of Space Studies, and at other labs here and abroad have all run greenhouse cases. The data from these models supported the prediction of a warming that could range from about 3°F to 9°F. The numbers vary as widely as they do for the usual reasons that distinguish one model from the next—whether the models are three dimensional; how good or bad their accounting for clouds might be; how fine a grid they use, and so on—but most of the modelers seem comfortable saying that a world with twice the current amount of carbon dioxide in its atmosphere will be around 5°F warmer than now. Allowing for the effects of the four trace gases, the models also provide a rough date for a warming at this scale: the equivalent impact of the modeled greenhouse on the real world could appear as early as 2030.

There is a catch, though: the models cannot predict with any degree of certainty the specifics of likely changes in the climate as a whole. Once they generate the increase in temperature, all the models indicate a chain of effects on the weather, but each model behaves differently, so there is no way to tell which of the new rainfall patterns, say, will resemble the one that will occur in this world. But the models do allow us to say that rainfall patterns will change, that some areas will become dryer than they are now and some, wetter. The most that a modeler can say is, "Be scared, because within our children's lifetimes the Midwest *may* dry up and the farms *may* blow away," or "Rejoice, because in our children's lifetimes the Sahel *may* see the rains return." Both of those changes have been predicted by at least one model.

James Hansen, the head of the climate research group at the Goddard Institute for Space Studies, has come up with a more concrete set of projections for a problem simpler than the rainfall question: how warm it would get in the actual places in which people live. Hansen woke up a rather sleepy congressional hearing in mid-1986 by running down the increases in the number of days with highs over 90°F certain American cities would have to endure in a world with doubled carbon dioxide. Washington, D.C., now puts up with 35 such scorchers each year; in the brave new world, that number goes up to 85. New York gets 48, up from 15; Memphis goes from 65 to 145; and Dallas passes beyond the realm of the livable, with a total of 162 days over 90°F each year, up from an already rough 100.

Hansen's testimony made for good theater and hence good politics. Such stories are almost fictions, for while Hansen's model did generate them, the results for a given place are far less certain than any calculation of globally averaged change. But the models are built to generate such tales. A three-dimensional model plays out its changes on a map on which one can find Washington, D.C., or Malacca, or Tombouctou. And ultimately, this storytelling power is crucial for our purposes because it is not global averages that matter, that are even comprehensible as daily experience: these fictions are what make the reality of climate change believable.

Another limit past which the models cannot go, even as guides, is to tell us how ecosystems will respond to a warmer environment; the models cannot conclude which species will benefit, which peoples will gain and which will starve. The interaction between climate change and local weather, between local weather and local crops and wild ecosystems, between the people of a village or a city and the landscapes that surround them—all of these are vanishingly too complex to simulate, and hence any modifications here are formally unpredictable. But it is, in fact, what climate change means to me, living in one place, now—it is this that I want to know. And so it is that I find myself standing on the railway embankment with John Teal, staring seaward across a marsh, seeking in a single patch of damp ground some link to the whole world.

The central fact of life here is the presence of the sea and salt. On the landward edge of the Sippewisset salt marsh, above the usual level of tidal floods but within range of salt spray, grows a particular mix of plants: poison ivy, bayberries, junipers. These plants create a ring of shrubbery around the edge of the marsh, but with even the slightest bit more exposure to salt they would die. Just downslope from them, about twenty feet or so off the railway embankment, the marsh proper begins. Here the community shifts to favor plants that can stand having their roots flooded occasionally by saltwater: marsh elder, for example, and goldenrod and marsh lavender. These become inundated perhaps twice a month, at the extreme high tides. Throughout the marsh little hillocks pop up—glacial till, the rocky fragments left behind when the last ice age retreated. The tallest of them stands no more than five feet above the level of the marsh, but they support little ecological islands of their own. Teal points out one near the edge of the marsh covered with patches of poison ivy, some pickle plants, and a juniper.

Down in the marsh proper, grasses dominate. Black rushes stand on the inland side, and then as the marsh slopes imperceptibly down, the rushes get shouldered aside by the classic northern marsh grasses: *Spartina patens*, a fine, fairly short, triangular plant known locally as marsh hay, and *Spartina alterniflora*, known as cordgrass. The plants' success is predicated on their ability to survive constant, almost overwhelming stress. Throughout the marsh, the twice-daily flow of saltwater ensures that conditions will remain on the very edge of what these plants can tolerate. If that flow alters, if for any reason the tides reach higher or lower than they do now, then the marsh must respond.

The present marsh is the product of a process that began with the end of the last ice ages. As the ice melted, the land on which the ice had rested rose, rebounding from the effects of the weight of the glaciers. Relative sea level therefore dropped, and marshes formed on coasts that stretched several miles out into what is now the ocean. Relics of earlier dune-marsh systems exist all the way up and down the east coast of the United States, submerged a few miles down the continental shelf from the current shoreline. As meltwater from the ice flowed into the world's oceans, the sea level began to rise again. Saltwater flowed into and over coastal flats, killing the old marshes and then inundating them. Finally, as the new shoreline migrated inland toward its present location, the tidal regions repopulated, and the marshes reestablished themselves. Spartina grasses invaded upslope, taking over ground that used to hold less salt-tolerant species. Those plants in turn migrated inland, pushed back by the pressure of incoming salty water.

As sea level rise determines where the marshes begin and end, it also, of course, can kill those that stand in its way (or those that a drop in sea level leaves drying out). An increase or decrease in altitude of perhaps one or two centimeters brings with it a very different mix of plants in the salt marsh; a flat that supports marsh hay, for example, can stand just that much above the level that supports only spartina cordgrass. If the sea level rises too quickly, then the marsh cannot keep up; it will die, drowned or simply poisoned by salt. How fast is too fast is hard to determine, but Teal guesses that the marsh could stand a rise in sea level of perhaps a centimeter or so each year. Two centimeters a year for three years would probably kill the marsh plants; five centimeters a year would certainly do so. The underlying fact of life in the marsh is that it is an extremely sensitive ecological barometer of changes in its surrounding environment.

The indications from this particular instrument are grim. As Teal and I stop by a small juniper bush, it is clear that the Sippewisset salt marsh must die. One of the easiest predictions to make of the greenhouse effect is that it will produce a rapid rise in sea level. The increase in temperature will, of course, warm the oceans as well as the land; as they warm, the oceans will expand. One model calculated that the oceans' heights would rise by 80 centimeters as they reached a new thermodynamical equilibrium with the warmer greenhouse world. To that rise, add the melt-water from the great continental glaciers that remain. Although no one has yet devised a satisfactory model of ice sheet dynamics, and hence there is no reliable prediction about the amount of melting that will take place, climate scientists remain worried about the fate of the west Antarctic ice sheet, which rests on bedrock that lies below sea level. If it breaks up, sea levels could rise by 3 meters or more.

Under any circumstances, farewell to the Sippewisset. If the rise of a single meter or so that seems to be inevitable happened slowly enough, under ideal circumstances the Sippewisset marsh would simply move back onto the Cape. But human intervention has made adaptation to human-induced climate change impossible here. The railway embankment at the back of the marsh stands only a few feet high, but it might as well be Hadrian's Wall. The marsh cannot pass it, and so as the sea level rises the spartina grasses will drown, and the salt will poison the rest. In time the marsh will turn into a shallow, muddy lagoon.

That is a loss in itself, of course. The marsh is beautiful if you happen to have Teal's eyes through which to observe it, with its subtle and complex variation of land and plant cover taking on the texture and capacity to surprise of a Turner painting. And it is an ecological loss of substantial proportions, if we take into account not just this marsh but the fate of the entire coastal wetland system. At Sippewisset alone, the enormous biological productivity of the marsh system supports, at one or another phase of their life cycles, shrimp, winter flounder, striped bass, herring, mullet, eels, and bluefish, among others. The disappearance of the marsh certainly won't help those species, though again, any guess about the direct impact of flooding the marsh on a given fish population is only that—a guess.

All this is bad enough. Following John Donne, I hold that any death diminishes the whole, even a death here of a few acres of muddy ground. But accompanying the hard fact of the destruction of an obscure piece of land is another cause for concern, one that reaches far beyond this corner of Cape Cod. Teal's certainty that the Sippewisset will disappear is based on the coincidence of two facts: the greenhouse warming and the attendant sea level rise, combined with the presence of the old railway cutting. A human structure makes it impossible for a natural system to respond to a human-created climate change. And while the Sippewisset's fate is easy to foresee, given its scale and the stark fact of that little railroad cutting, it provides in microcosm an image of disasters on a far larger scale. In Bangladesh eight million people have built homes and farms and roads on land that floods occasionally now and that will be uninhabitable should the sea level rise one meter. If it rises three meters, twenty-five million people will be displaced.

These people have no feasible way to adapt to the change in conditions: they have made their lives there, own land that they cannot carry with them, have built their homes there. People in coastal areas the world over will have to accommodate to an ocean that reaches higher than they have ever seen. Venice will have its present difficulties multiplied; Holland will have to look to its dikes; Galveston might not make it past the next killer hurricane, should a great storm strike at a time when the tides are higher than they ever were during earlier blows. Cities, railroads, farms in the shadow of a coastline armored against the sea—all the structures we have built limit our ability to bend to change, just as the Sippewisset is frozen in place by that little embankment. What this means is that as the greenhouse effect begins to sweep through the environment, we will, as matters now stand, simply have to take the punishment.

Writ large, on a global scale, the easiest message to read is simply that large numbers of people living along coastlines will probably suffer—loss of property, of livelihood, of life. That scenario creates its own imperative for action. But another message presents itself, and it is seen easily from where Teal and I stand on the edge of the marsh. It is that the problem isn't simply climate change, or even climate change as it encounters human artifacts; the disasters that impend are the results of beliefs about the dynamics of nature that are false. When the railroad company built its line along Buzzards Bay, it almost certainly had no concern for climate, except to assume that the weather would stay more or less the same, more or less indefinitely. People everywhere make that assumption all the time, but the Sippewisset is doomed because of such faith. There is no escape hatch, nothing built in to allow for flexibility in the face of change.

This is how the greenhouse effect will produce its most subtle and lasting challenge: systems that remain sufficiently malleable to bend under the impact of a global warming will probably endure, and the ubiquitousness, the fundamentally inescapable nature of a wholesale change within the atmosphere, will expose those systems that do not possess the necessary abilities to adapt. Natural systems, on their own, do exhibit some degree of this kind of flexibility. Again, the salt marsh provides a model that describes how the greenhouse world might behave. The Sippewisset lives on stress. Different tolerances for salt create a precise and elaborate structure within the ecosystem, and every fluctuation in that salt stress ripples throughout the marsh, enabling a woody shrub to grow, for instance, in an area grasses had dominated or sending the grasses on the march should the salt advance. Every fluctuation alters the mix of plants on that particular piece of land. Shifts in the mix mold the behavior of the landscape as a whole. A changeover from one species of spartina to another as the dominant grass of a marsh is reflected in the biological productivity of the marsh—*Spartina patens* is far less productive than *Spartina alterniflora*, and marshes that it dominates support less marine life than ones in which the *alterniflora* species thrives.

At the global level, the greenhouse effect will cause a wholesale turnover in the kinds of stresses ecosystems everywhere experience. The change in temperature will have a direct impact. Boreal forests probably will behave in a manner directly analogous to that of a marsh moving upslope from an advancing ocean, pushing north into areas that now support only the shrubs and grasses that stretch into the tundra. Maize may flourish where wheat grew before, and the grain belt as a whole may move into the higher latitudes.

Or it may not. Rainfall shifts will affect agriculture, though in the absence of any reliable models of those changes it is impossible to say which area will benefit by a sudden easing of climate constraints on its farm output and which will suffer as the heat builds and the rains do not fall. And ecosystems themselves do not move as units. Some species may migrate and thrive under new conditions, while others will not. Whatever the end result, the new communities of plants thrown together by climate change will be different from the ones we know now, and some species, almost certainly, will disappear.

It gets worse, or at least harder to read. Carbon dioxide acts as a fertilizer. If the concentration of carbon dioxide in the atmosphere increases, the two main classes of plants (distinguished by the different mechanisms that they use to fix energy from the sun in photosynthesis) both benefit, each fixing more carbon dioxide for each unit of sunlight. One class, called C3, gains more—a member of the oleander family, for example, a C3 group, would increase its photosynthetic rate by about one-half if the carbon dioxide levels of its atmosphere doubled over that of the present day, whereas *Zea mays*, a C4 plant, would improve by just one-fourth. The general increase in photosynthesis could be good news, particularly if it raised the productivity of agriculture. But it surely means that the mix of plants in any given ecosystem will change, as the variable photosynthetic rates would enable some plants to expand their presence and force others into a narrower and narrower niche. Such changes from a human perspective could be good, bad, or possibly neutral. The point, again, is that the world will be different, and we will have to be prepared to adapt to the change.

In the Sippewisset in October, the plants are doing relatively poorly. The grasses throughout most of the marsh are brown and seem unenthusiastic. Down along the creeks, however, each bank is lined with stands of grass that remain green and stand taller than the surrounding field. The creeks carry nutrients and oxygen with which to oxidize those nutrients, providing a source of fertilizer for the plants nearest the moving water. It is the luck of the draw for each individual: the ones that happen to grow closer to the creeks grow better, more vigorously, than members of the same species in another place. It is a natural process, an ordinary one, and one that will continue; throughout nature it is possible to find, if one keeps score, winners and losers. The importance of the greenhouse effect is that it changes the shape of the global playing field. From the level of individual plants, to whole ecosystems—and nations—there will be a new set of winners and losers.

This idea is an anthropomorphic one: we impose our notions of victory and defeat on the daily reality of nature, where such concepts may make little sense. But they do matter to people. Some towns and farms, perhaps whole nations, will benefit from a global change in atmospheric chemistry. But others will not; poor Bangladesh, facing the specter of a rise in sea level, is an obvious candidate for a loser's mantle. More important, most of these creeping disasters or bouts of good fortune will not be obvious—

since it is simply impossible to gauge how the biosphere will react when carbon dioxide levels double, temperatures rise, and rainfall patterns shift.

And such an inability to anticipate creates the greatest risk. The specific impact of the greenhouse effect is unknowable, at least at this point. Some people will starve when rain doesn't fall on their farms. If their neighbors over the mountains are doing better, they may choose to move, and if the neighbors don't appreciate the additional burden of refugees or claimants on their land—well, wars have started that way in the past.

To this concern one may plausibly respond: "Don't just stand there—do something!" The possibility exists that human actions now could undo, or at least mitigate, the effects of a process already under way. With foresight one could imagine replanting forests on a large enough scale to fix some of the carbon dioxide that has built up in the atmosphere. Prudence would dictate that people living in low-lying coastal regions begin to think about the protection they will require over the next couple of generations. We could filter carbon dioxide and the other trace gases as they are being produced since it is at least technically feasible to capture the carbon dioxide being vented from power stations before the exhaust plume reaches the open atmosphere.

In fact, while reforestation is probably a good idea on its own merits, it probably cannot do much to slow the accretion of atmospheric carbon dioxide. As a United Nations' Environment Programme research paper noted, "A much larger area than France would have to be reforested annually to 'mop up' the carbon dioxide from fossil fuel burning." Building a system of dikes to keep out the sea is easier to imagine—but as the same U.N. document points out, it will be far more difficult for a Bangladesh to prepare for the change than it will be for a nation in the developed world. Finally, while it is technically feasible to fit exhaust streams with filters, it is extremely expensive, and to date no one has made any real effort to do so for carbon dioxide or any of the other trace gases.

That leaves only one option: Cut off or cut back production of the trace gases at their sources. If you fear the consequences of the buildup of greenhouse gases, don't do the things that produce them. What that means in practice is to burn less fuel: conserve oil and coal; use oil, which releases less carbon dioxide for each unit of energy used than coal, wherever one has the choice; replace both with other fuel sources; choose to live in a way that requires less energy per person—in all, change the way people in the developed and developing world live, nation by nation, across the planet. Certain steps would involve less pain than others. Since natural gas produces 60 percent of the carbon dioxide that coal does, the latter should be replaced by the former whenever possible.

The very best that could be achieved here is a slowing of the pace of climate change: After all, leaving aside the question of whether or not long-term reductions in per capita energy use are likely, the other trace gases also matter. Nitrogen oxides and methane pose a special problem, for unlike carbon dioxide, they are harder to track to specific sources. But the core issue remains simply whether human society today will make any effort at all to ward off this climate change until people have a chance to adapt to it. We are faced now with a test case, a kind of trial run, that will allow us to discover within the next few years whether we as a species can act to give ourselves that breathing room.

The test involves the juxtaposition of the greenhouse effect with another perturbation to the global climate machine: the threat of a reduction in the stratospheric ozone layer. Ozone is produced in the stratosphere by a series of reactions that are powered by ultraviolet light from the sun. When ultraviolet light strikes molecules of oxygen, it splits apart its two component atoms. Each individual atom is then free to bind with another oxygen molecule, forming the triatomic compound ozone. The proportion of ozone relative to molecular oxygen remains roughly balanced because ultraviolet light splits ozone molecules as easily as it does the more common oxygen molecules. This photochemical dance in the stratosphere has a practical consequence: a substantial amount of the ultraviolet light that would otherwise strike the surface of the earth is instead absorbed by the energy requirements of stratospheric chemistry. Because overdoses of ultraviolet light can cause everything from sunburn to melanoma in people and genetic changes in at least certain plants on land and phytoplankton in water, we have a substantial stake in retaining the protection of the ozone layer.

In laboratory experiments in 1974, Sherwood Rowland and

Mario Molina, two researchers at the University of California at Irvine, demonstrated how that ozone layer could be at risk. Ozone is peculiarly vulnerable to a set of reactions involving the element chlorine. Chlorine can attack a molecule of ozone, capturing one of its three oxygen atoms to form chlorine monoxide; when that molecule encounters another single atom of oxygen, another reaction occurs, producing one molecule of ordinary molecular oxygen and one atom of chlorine, which is then free to break up another ozone molecule. Although chlorofluorocarbons (CFCs) contain chlorine, at the surface of the earth they are quite stable. Once they enter the atmosphere, though, they drift slowly upward, toward the stratosphere. Rowland and Molina showed that when CFC 11 and CFC 12 are exposed to ultraviolet light, they break up and release atomic chlorine; in their experiments the two researchers proved that at least under laboratory conditions, CFCs were quite effective in destroying ozone. If CFCs in the stratosphere behaved similarly, they warned, and if CFC production continued to increase in line with the trends established then, between 7 and 13 percent of the ozone layer could be destroyed within a century.

At the time the U.S. government responded by banning the "nonessential" use of CFCs in aerosol cans, and public fears for the ozone layer were stilled for more than a decade. But worldwide use of the substances has continued to increase, for, apart from the problem of ozone, CFCs are wonder chemicals, stable, nontoxic, and extremely useful. Any city that lives and dies by the air conditioner runs on CFCs; they are used to blow the foam that produces Styrofoam; every home with a freezer or a refrigerator needs them. CFCs have become a multibillion-dollar industry. Between 1974 and 1986 the major producers of CFCs within the United States argued through their trade organizations that scientific evidence of a connection between CFC production and a threat to the actual ozone layer (rather than Rowland and Molina's laboratory re-creations of it) had simply not been established: no one could demonstrate that the ozone layer had actually been depleted. Hence, they concluded there was no need for any regulation of CFC production beyond the aerosol can ban. The U.S. Environmental Protection Agency agreed.

Then, in 1986, a group of British scientists in Antarctica reported a strange anomaly. They had been measuring ozone levels above that continent since 1957, and over the last five years they had noticed a sudden drop to about half the normal values in the amount of the gas. The hole in the ozone layer lasted each year for about a month and a half, beginning in the early Antarctic spring (toward the end of September, about a month after the Antarctic sunrise), deepening during October, and then disappearing quickly in the first several days of November.

That report launched the modern phase of the ozone debate. In August 1986, the National Science Foundation (NSF) sent a team of researchers to Antarctica in an attempt to identify the process responsible for creating the hole. One proposed explanation invoked physics—changes in wind patterns that could affect Antarctica exclusively. Ozone is produced most vigorously in the tropics, and planetary winds can carry ozone-rich air from the tropical stratosphere south to Antarctica. In model experiments, a weakening of that wind pattern can lead to the ozone depletion, but so far a mechanism that could cause such a weakening has not been found.

The other alternative came from chemistry. It was suggested that chlorine from CFCs might accumulate over Antarctica and lie dormant until the sun rises and provides the necessary energy to power the reactions that consume ozone. Unfortunately, the researchers in the 1986 NSF mission were unable to measure the presence of chlorine monoxide directly within the hole, which would have essentially nailed the issue down. One of their instruments did detect chlorine monoxide at or near the top of the hole, while another found $OClO$ (chlorine bonded to two oxygen atoms) within it. Neither of these discoveries is the smoking gun that proves CFCs are to blame. The scientists reporting the chlorine monoxide measurements would allow themselves to state in the pages of the journal *Nature* only that their findings "leave little doubt that the development of the Antarctic ozone hole involves chlorine chemistry as a direct and prominent feature." That's one step short of fingering CFCs as the culprit, but in a return visit the following year further supporting evidence was found that chlorine from man-made sources was present in the ozone hole, driving the process of depletion further. The most recent measurements of the hole itself also show, most ominously, that it is getting larger and is lasting longer.

Yet though the evidence appears to be coalescing, implicating CFCs ever more in the creation of the ozone hole, moving from Antarctica to the planet as a whole is extremely difficult. The hole is large, dramatic, incontrovertibly there: scientists can go and point their instruments at it every year until they find a solution. Tracking the more subtle declines in average ozone concentrations worldwide is a much more complicated problem. NASA has tried to monitor global ozone concentrations using two satellites. One of them has detected a 5 percent drop over six years, but the satellites themselves are not perfectly reliable. Instruments in space drift over time, and six years is long enough to make people very nervous about the results from any instrument they cannot reliably calibrate. Again, all that can be claimed is that CFCs are the most likely cause of the hole in the ozone layer over Antarctica. They seem to be triggering a global decline in stratospheric ozone concentrations, but as of now there is no absolute proof that they are.

In the past, without proof more solid than now exists, CFC producers have resisted any suggestion that controlling CFC production might be a good idea. But, in what might turn out to be a major break with old habits, in the fall of 1986 the wall cracked. The American CFC trade organization agreed that prudence required, in the absence of certain knowledge about the possible long-term effects of CFCs on the atmosphere, that any increase in CFC production should be controlled. At the same time, the U.S. government proposed an international agreement that would freeze CFC production throughout the world now and would phase out altogether certain CFCs within ten years; after initial resistance, the major CFC-producing nations signed the Montreal Protocols, which freeze and then limit CFC production. Finally, perhaps most remarkably, in 1988 Dupont, the largest producer of the compounds in the United States, announced that it was phasing out production of CFC 11 and 12 altogether.

Dupont's action establishes a crucial precedent. It comes in the context of reports that the ozone layer may be depleting more rapidly than thought and at a time when, even if all CFC producion stopped now, the backlog of chlorine in the atmosphere is ufficient to cause considerable further depletion before all the chlorine is swept from the upper atmosphere. But even if the perfect outcome—a reversal of the damage already done—is not mmediately forthcoming, the example set for confronting other global issues is extraordinarily valuable. It argues that if we don't know what will happen and if the possible damage is great, then we must slow down, stop. We must not even open the possibility of altering fundamental processes that have evolved over aeons; the odds are good that if we succeed, we will not enjoy the results. That he threat of ecological damage is enough to influence policy, nstead of policymakers requiring some absolute standard of proof, s an important step forward.

And it is possible to be optimistic that the rest of the world will follow Dupont's lead. As global ecological issues go, CFCs provide one of the relatively easy ones. Though they are extraordinarily useful chemicals, substitutes are available; CFC 22, for example, is far less destructive of ozone than CFC 11 and CFC 12, and a number of companies have been developing other compounds hat could serve. In the great scale of things, if it meant saving our ozone layer, we could surely find alternatives to Styrofoam in which to pack our burgers and our coffee. It helps that only a imited number of sources, basically large chemical companies, produce CFCs. We know who they are and where they are, so if he international community agrees to halt all production and emissions, it won't be that difficult to enforce the ban. And finally, such a ban would yield an added global benefit: not only do we protect the ozone layer against a possible threat but we slow the greenhouse effect. If any international environmental initiative is possible, it should be this one.

And it will be our success or failure on this issue—success here defined as control and regulation of our actions until we know better what we may be doing—that will reveal whether there is any hope at all on the larger issues to come. If we cannot control CFCs now, we will almost certainly fail to slow the onrush of the greenhouse effect. On the broadest level, failure to act now, on an issue for which the risks and benefits of action seem clear, would to me presage folly after folly. If we cannot accept this responsibility, then we cannot expect in the future to be able to protect ourselves from anything we might come up with that could threaten some part of our climate or the ecosystems that support us.

Acid rain, desertification, the greenhouse effect—these are obviously distinct phenomena, occurring on different scales of space and time. But a line of connection stretches between them; together they illustrate a fundamental flaw in the relationship between human society, in the mass, and that interaction of land and air and water that collectively produces the conditions we call climate.

The line is formed by what has been dubbed "the tragedy of the commons." The concept derives originally from landholding practices in preindustrial England. In the rural areas, the commons were patches of land held by the village as a whole; any villager could pasture livestock there or glean some fuel against the winter. Overgrazing could, of course, degrade the common land, but a village is a small enough unit to have some control over its members, enough, probably, to keep any individual abuses in check. Today the global climate system is, in one sense, a common resource—everywhere, everyone depends on the continued, more or less stable behavior of his environment. We expect a given temperature range, a certain mix of plants, a certain range of rainfall, and so on. But as human ability to produce change on a global scale grows with each passing year, these certainties are undermined. Individual acts—by single people, or cities, or nations—have global consequences, but despite any rhetoric about a global village, we lack any social structure cohesive enough to keep each of us in check.

The problem is partly that a sense of the commons itself has eroded over the last couple of centuries. The notion of a commons implies a relationship between the people and the resource held jointly. A stand of timber isn't simply inert, out there to be used; it is part of the community, maintaining it and to be maintained by it, to ensure the continued existence of both. We don't have that conception of our atmosphere. It exists outside of human society— a constant. We breathe it, fly through it, dump waste into it, use it—and it is simply there. The same is true of the oceans, of rain forests, of groundwater, and so on. The natural world is an arena in which our individual aspirations, as people or nations, have had unlimited play.

The revelation of the new science of climate is that this will not do, not forever. The structure of thought built into the methods of that science contains, indeed derives from, the idea of connection, of interdependence, of reactions following on actions within the web that binds human society with those inconstant systems, the atmosphere and oceans—climate. That's the point of a general circulation model—it is a device designed to simulate the action of the whole planet, human beings and their environment in concert. To a great extent we see what we look for, and with these models, with the satellites peering down from above, circling the world fourteen times a day, what becomes clear is that a need for a renewed commitment to the commons is inescapable. The atmosphere knows no borders; it is vulnerable to human action; it is changing now. That much, at least, is scientific fact.

That ought to be enough to arouse a sense of enlightened self-interest, along the lines of villagers keeping tabs on anyone pasturing an extra sheep or two on the village green. But we can see something more than a picture of the climate if we look closely enough at the modeled greenhouse worlds, or at our own world through a satellite's eye. In these images resides a kind of self-portrait, for the changes our new tools record are measures of ourselves and our capacities. Within that picture we can find a measure of our quality, our moral progress. The science by itself makes no moral claims, of course, but ideally it informs such claims. Mine is almost a gut reaction: we revile the bird that fouls its own nest. Clearly we now have the power to soil ours; having that power and being aware of it carry the imperative to behave properly, prudently. To do otherwise would be to do wrong because we know better now.

Call Me Ishmael is at once Charles Olson's meditation on Melville and on the character of Americans. In it he wrote, "We are the last 'first' people. We forget that. We act big, misuse the land, ourselves." We do, we do indeed, and in this century, at this end of a very long century, it is not just Americans who play the part of "first peoples." We are new to this world, new in our numbers, new in our capacities to produce and destroy, new in our power to monitor the implications of both. We act on the grandest scale; we misuse the land—and air and water—and in so doing we diminish ourselves.

The Sippewisset salt marsh is a brown, flat, small piece of land that virtually no one on earth will ever see or miss. I will miss it, now having had the good fortune to see it through eyes that knew how to appreciate it. From where I stand, looking out toward the line of dunes that block the sea, I can just make out something odd. Along the dune crest runs a kind of jagged line. It is, Teal explains, the remains of old Christmas trees, stuck in the sand in a more or less futile attempt to prevent the dunes from washing away. The dunes here have existed for millennia without benefit of a fence of pine rubbish, but jetties and docks up-current now trap the sand that used to flow onto the beach, and each succeeding storm erodes the shoreline further. The sea may rise, or the dunes may fall; either way the Sippewisset will be gone in a short time. That sounds a warning for the great world beyond the marsh. It is not, necessarily, a prophecy.

CHAPTER 13

Tales of the Future

GALILEO, HOLDING that corrosively novel tool, the telescope; Galileo, facing his inquisitors, becomes one of the great symbols of the modern age, of modern thought. Galileo took his new tool and looked outward; in every part of the sky at which he aimed it he saw wonders, miracles to him and to his fellows. But Galileo's telescope could not remain fixed only on the night sky; it twisted in his hand, in the hands of his age, and pointed inward, toward an unscientific destination, his soul, toward the core of his time. The myth has it that when Galileo confessed his errors and assented to the orthodox claim that the earth stood stationary while the sun revolved around it, he turned away from his inquisitors and mumbled, not quite under his breath, "But still it moves." Within that myth, this truth: the earth moves.

The fact of motion and the knowledge of movement transformed Galileo's world, in time. What endures in legend as a symbol, though, is not the specific discoveries; they have become commonplaces, so unremarkable that today it is almost impossible to imagine how the world would appear through sixteenth-century eyes or how to believe, as the Church instructed, that the earth rested at the center of creation. What endures is that impossibility and the restless knowledge that every increment of discovery can change the world, remake what we see beyond us, and how we see ourselves within a world of constant change.

A story from the recent history of climate science captures the essence of how science performs this twin act, simultaneously transforming our relationship to the world and recasting it again within the context of the new perspective. It involves a feat of pure imagination, an apocalyptic vision played out on a world that does not exist; it is the story, thus, of the conscious effort to produce a myth that could command belief, and with belief, action.

The story begins with almost a stray thought, a question that nibbled at the edge of the questioner's mind until it became impossible to ignore. In 1981 and 1982 the Swedish journal *Ambio* sponsored a multidisciplinary study of the long-term consequences of nuclear war. The editors of the journal asked Paul Crutzen, a Dutch scientist working in Germany, to repeat and extend research documented in a 1975 National Academy of Sciences report that suggested a novel mechanism of global apocalypse—the idea that a major nuclear war could sufficiently disrupt the atmosphere of the earth to ultimately threaten the survival of all life on the planet. In the National Academy scenario, fireballs from nuclear explosions would inject large quantities of nitrogen oxides into the stratosphere where those compounds could destroy ozone. Crutzen was asked to update the atmospheric chemistry in the earlier study and to produce a scientifically credible scenario of the effects on the stratosphere of a nuclear holocaust.

One of the central assumptions of the 1975 study was that a nuclear war would involve a large number of very big explosions, with both sides using warheads of one megaton or more. (A megaton explosion is equivalent to detonating one million tons of TNT; a kiloton corresponds to 1,000 tons.) However, by 1981, most of the warheads in both superpower arsenals had a much smaller yield—America's Minuteman missile, for example, carries three warheads each with a yield of about one-third of a megaton. When Crutzen and John Birks, a colleague from the University of Colorado, examined a nuclear war scenario that consumed larger numbers of smaller warheads, they found that much of the nitrogen oxides did not reach the stratosphere, which meant that the ozone layer appeared to be significantly less threatened than previously thought.

The last thing that Crutzen wished to do was to prepare a paper that even seemed to suggest that nuclear war might not be so bad after all. So the two scientists went back to look for other damaging atmospheric effects. They suggested, for example, that the smaller warheads would produce a large amount of photochemical smog near the surface of the earth, creating a kind of nuclear Los Angeles wherever the bombs fell. Then, within a few weeks of their deadline, Crutzen was struck by a sudden realization: where there is fire, there is smoke; where there is smoke, there is a shadow; and in the shade, photochemistry ceases and plants die. Crutzen and Birks swiftly did a set of simple calculations on the assumption that a million square kilometers of woodlands would burn in the nuclear conflagrations, releasing 400 million tons of soot into the atmosphere. They concluded that that cloud of smoke could block 99 percent of the sun's light from reaching the surface of the earth for as long as several weeks.

Such an event would, of course, be a disaster. Crutzen and Birks had performed only preliminary calculations, but it was clear even from a first pass that if they were even close to right, then nuclear war had suddenly become vastly more horrible than the horror it already was; it threatened, with its prolonged period of darkness, the life of every plant on land and on the surface of the sea. It was not, this result suggested, just the combatant nations who were at risk; everybody was.

In itself the idea of universal annihilation isn't new. In fact, a number of scenarios had been proposed about the ways in which nuclear war could destroy the world. In 1945 the fear was that a single nuclear weapon could ignite the entire atmosphere; in the 1950s people believed that nuclear war might produce enough radioactive fallout to kill all life; in the 1970s, as the 1975 study suggested, the foreseen threat involved the ozone layer. Each of these apocalyptic visions was discredited in time, however. They all posed the prospect of some chain of destruction triggered directly by the initial explosions, but by the 1980s, it appeared as though the direct effects of the bomb—blast, heat, and prompt radio-activity—lethal as they may be, were essentially local and regional phenomena. But lots of bombs could kill lots of people: it has been estimated that a full-scale nuclear war involving a large portion of the global arsenal would kill a billion people outright, with another billion likely to die more slowly from the effects of radiation, disease, and starvation triggered by the social disruption attendant on such a war. Even in the worst case, however, it appeared as if it were impossible to destroy the entire human world.

For the first time, then, Crutzen and Birks had proposed a plausible mechanism by which the bomb could cause greater damage through long-term, indirect effects than through the immediate consequences of the original blast. Their study also reaffirmed that the global climate machine was vulnerable to human action, whether leisurely changes like the greenhouse effect or a single, swift, catastrophic blow.

Within climate science, Crutzen and Birks's result crystallized a view that had slowly been forming from results in apparently unconnected lines of inquiry. There are enormous dust storms on Mars, and when one occurred at the time of the Mariner 9 mission in 1971, instruments on board the lander observed that beneath the shadow of the storm the planet cooled. At NASA's Ames Research Center, James Pollack and Brian Toon, along with Carl Sagan of Cornell, formed one of the groups that tried to calculate in simple models the effect of dust on planetary atmospheres and temperatures generally, including those of the earth. Then in 1979 the Alvarezes came up with their theory that a collision with some extraterrestrial object produced debris that could have cooled the earth long enough to cause the great extinction of 65 million years ago. Richard Turco, a researcher at a California defense think tank, joined with the NASA researchers to model that event.

So by early 1982 it was known that immense clouds of dust circulating in the stratosphere could, in theory, at least, affect global temperatures. When Crutzen and Birks's paper first began circulating in the research community, the idea that smoke would have a more significant role to play than dust in the postnuclear climate struck an amazingly resonant chord. Turco used his own model to try to estimate how much smoke the fires from burning cities would produce in a variety of nuclear-war scenarios. Then he calculated the impact that that much dark, heat-absorbing material in the atmosphere would have on temperatures on earth. In collaboration now with four other scientists—Toon, Pollack, and Sagan, joined by another NASA researcher, Thomas Ackerman—Turco found that for a reference simulation of the aftermath of a major nuclear war, mean annual temperatures would drop across the Northern Hemisphere by as much as 35°C (63°F), and abnormally cold temperatures might persist for over a year. Any survivors of a nuclear war could, according to this first attempt at simulation (known as the TTAPS model, after the initials of the five scientists), simply die more slowly, shivering in the dark.

This, at least, was the picture painted by Carl Sagan at a public conference on the phenomenon, now dubbed "nuclear winter." The meeting was held on Halloween in 1983, and the climate modelers were joined by biologists who had attempted to assess the threat to life on earth posed by the climatic effects envisioned by the TTAPS model. The combination of cold and dark would be sufficient, particularly for a spring or summertime war, to disrupt the metabolism of virtually every plant; the sudden shock would cause significant dieback and possibly extinctions that would ripple through the food chain, thus threatening, perhaps, every species of animal trapped beneath the pall. In the worst case, according to Paul Ehrlich, rapporteur for the biological study group, "We could not exclude the possibility of a full-scale nuclear war entraining the extinction of *Homo sapiens*."

I was at that conference and heard Ehrlich's remarks. I remember the real sense of terror that they inspired—that they were meant to inspire. His was a tale of the end of days, and the landscape he evoked for me was worthy of Hieronymus Bosch: I could see the scourge of hellfire and the agony of the moment of attack compounded (in the side panels of the tryptich) by the lasting, ongoing torment suffered by those who lingered on in a world diminishing into nothing.

This was a conscious act of mythmaking, "myth" here in the sense of a story told to concentrate experience. Ehrlich and Sagan and the rest were not spinning fables, not at all; they meant to compose a plausible, easily understood portrait of the world as it might be that would act as a lever with which to move the world as it is. They wished to evoke fear, and belief, and then response.

This they did through an exercise in climate science. There was, in the genesis of the idea of nuclear winter, a kind of epiphany in the field, a coming together of the disparate strands of research that have together altered the view we are able to take of how this planet works. Ideas do not burst forth whole, like Athena from Zeus's forehead; this one, at least, grew out of ground that had been well prepared. The dinosaur extinction problem is a study in holocaust, and the proposed solution—the catastrophic collision with an asteroid or a comet and the cascade of disaster that followed in its wake—gave rise to essentially the same concept that underlies nuclear winter. Crutzen himself began to study smoke after George Woodwell had suggested that the burning of tropical rain forests was a major source of carbon dioxide in the atmosphere. To test the idea, Crutzen went down to Brazil in the late 1970s to collect smoke samples to actually measure the CO_2 content of the plume. In so doing he found that Woodwell had probably overestimated the amount of carbon that deforestation would release. More important though, Crutzen began to gain a kind of "feel" for smoke, a sense of its importance. Martian dust storms, volcanic eruptions here on earth (great eruptions, like Tambora in 1815 or El Chichon in 1983, which spew out enough dust to cool the earth a little), El Niños, which provide direct experience of the dynamics of heat exchange between the atmosphere and the oceans; acid rain and the fallout from above-ground nuclear tests, which provide models of global transport of pollutants: the list goes on and on, but all of these events are relevant to the question of what will happen to global climate after a major nuclear war. Without all these lines of research—without the direct experience of climate change (like living through a major El Niño, for example) and the slowly gathered knowledge that has uncovered the links between place and place, system and system, ocean and atmosphere and plants and ultimately human beings—without such discoveries it would be impossible to imagine the mechanism that could disrupt the climate system on a global scale.

And so, in one sense, nuclear winter is nothing much new. It is simply, like other exercises in science, an extension of research that has gone before. It was taken as both authoritative and credible because it so clearly echoed research that the scientific community had already accepted. What on its face is an amazing claim—that the actions taken on a single day could transform climate globally (or at least hemispherically) for months or more—seems less outlandish when proposed in a context of dying dinosaurs and a world in which a change in atmospheric pressure over the southern Pacific can trigger record rainfall in Louisiana half a year later.

In another sense, of course, nuclear winter is absolutely revolutionary. It is incontrovertibly an exercise in scientifically generated fiction; the nuclear-winter world is a made-up world, a make-believe world. It exists entirely within a handful of computer models, and all the model experiments include some leaven of "what ifs." But the nuclear-winter simulation is qualitatively different from more conventional exercises, such as those involved in tests of the greenhouse effect. There is no analogy to a full-scale nuclear war; there is only the war. Simulations of it are necessarily explorations of the possible, first, not necessarily of the plausible. Every such simulation is based on arbitrary choices—the size and number of warheads, which targets are hit, how completely they burn, and so on and on and on. It follows that claims for given outcomes of these wars, beyond the obvious one that the lives of enormous numbers of people will come to an end, are equally fictions.

Nevertheless, models fill a particular function: to discipline the imagination. They lend a certain authority to a given hypothesis that attaches specific numbers to specific assertions and force their authors to state, for example, how many tons of smoke they expect to see, how much of that smoke might rain out quickly, and how much might remain in the pall. And crucially, they provide the plot for each of these grim stories. A model is a world, with days and nights and winds and rain; with electronic smoke whirling around in electronic skies and with the passing of every electronic day, the events of nuclear winter proceed as in a story, a tale, a myth.

What was predicted and announced on Halloween is only a possible future for our world. It hasn't happened, obviously and thankfully, nor is it happening now: this isn't an experiment-in-progress in the manner that research on the buildup of carbon dioxide serves as a kind of global experiment in atmospheric physics and chemistry. While it uses the methods of climate science, analogies to historical climatology, and repeated model experiments, nuclear-winter research differs from conventional research in that it is unverifiable, until and unless we blow ourselves up.

Hence the myth. Nuclear winter is a story told to frighten us into finding some way to keep us from finding out—ever—if the prediction is right or wrong. Visions of the end of days are as old as human memory (come Gabriel and blow your horn), as is hubris, the pride that leads directly to a fall. Such fears animate the picture of nuclear winter, the destruction of the earth triggered directly by human folly. And we believe—we believed in 1983 on Halloween—because those old, familiar fears were recast within the context of a major research tradition, a rich vein of scientific discovery, all the novel findings of the science of climate that had emerged in the last decade.

To recognize the connection between ancient myth and modern science is not to criticize nuclear-winter research. Science ought to generate myths and cannot, in fact, ever keep from doing so. The existence of nuclear weapons begs interpretation, some effort to provide coherence and meaning. Nuclear winter is one of the results, an attempt to describe what the experience of a nuclear war would involve. Similarly, we speak of the greenhouse effect and illustrate it with a prediction of three months of 90°F heat in Washington, instead of its usual thirty-five scorching days each year. We tell ourselves stories to understand, to persuade, to force action, to alter or adapt to one part of our material world or another.

Greek drama features the device of the *deus ex machina*: the god who appears out of the machine to set things in order at the end of the play. The extraordinary power of science to explain, and to do so in a way that commands belief, creates a kind of *deus ex scientia*—the idea that knowledge alone, or further research, or simply telling the tale with enough authority, will rescue us, willy-nilly, from any difficulty in which we might find ourselves. The nuclear-winter case seemed, for a time, to compel a change in behavior that would have made the world infinitely safer.

One of the findings in the first study was of a threshold of safety, of about 100 megatons or so. If at least that much explosive power were detonated in an exchange, or even by just one side, then the smoke produced could still be enough to generate a cooling large and long enough to incur most of the disasters predicted for a much larger war. The only way out, Sagan suggested at the time and has argued since, is to reduce global nuclear armaments to a stockpile of some number of weapons of less than 100 megatons. As of this writing the United States and the USSR possess jointly somewhere between 15,000 and 20,000 megatons' worth of warheads.

For obvious reasons, we would share Sagan's wish to see as many nuclear weapons eliminated as possible. But Sagan has fallen into a trap. The myths generated by science are touched with a special quality that distinguishes them from the tales of another day. The glory of the older myths was in their certainty: If Odysseus's fleet scattered, it was because Poseidon willed it, and the message was do not anger the god if you can help it. When the God of the Jews spared Nineveh after Jonah's mission of prophecy, the message was behave well and God will—not might—spare you. But in science, climate science, the models do not afford so easy an equation. In the four years since the Halloween conference, additional research into nuclear winter has robbed us of the initial simplicity of the conclusion, and with it the meaning of the myth that the science still engenders.

The most recent research has focused on areas where it is possible with the existing data and models to reduce at least some of the uncertainty inherent in forecasts about the world after the war. The TTAPS modelers used a one-dimensional simulation in which a certain amount of smoke was injected into the model atmosphere and spread evenly over a planet that was all land or all ocean. After an arbitrary portion of the smoke (based on their best guess) was washed away by rain, they then calculated the temperatures that would result. In subsequent attempts, several climate-modeling groups used three-dimensional models in order to begin with a simulation that could reproduce many more of the features of real-world weather. They also began to modify other specialized models to generate the thunderstorms that would wash smoke out of the sky, and they tried, by making surveys of the burnable material available in cities, to gain a more accurate account than was available to the TTAPS team of how much smoke would actually be generated.

The results of these experiments, taken together, have led to a reduction in the claims for the severity of nuclear winter. In the most comprehensive recent study, Steve Schneider and Starley Thompson used their variant of the NCAR model to refine the original TTAPS picture. Their version included a mechanism to rain smoke out, and it produced patchy clouds of smoke that spread irregularly with the winds and that dissipated more quickly than the TTAPS equivalent. The patches meant that some areas were densely shadowed, which produced a new climate problem that they called "quick freezes"—areas that chilled rapidly to below

0°C beneath thick, local masses of smoke. They also found that their temperatures varied on large scales, with areas nearer the oceans cooling less than the middle of continents. For an average, the NCAR model indicated that, in a midsummer war, the temperature drop over the middle latitudes of the Northern Hemisphere would be about 12°C, which, when various adjustments have been applied to make the two model results more directly comparable, turns out to be about one-third as cold as the original findings suggested. Most important, while the larger the war the greater and more varied the meteorological consequences, the NCAR scientists could find no threshold, no magic number of megatons and fires, that would or would not trigger catastrophic climate effects.

The first thing to notice about this work is that while the Schneider-Thompson postnuclear world is not quite as horrible as the TTAPS world (they, half jokingly, call it "nuclear fall"), the climate effects they predict still generate unprecedented harm. A drop of 12°C during the growing season could destroy much of a year's crop across the middle latitudes of the Northern Hemisphere; the quick freezes could kill plants, as well as any animals and people weakened for any other reason, even if the cloud thinned and the frozen areas warmed within a week or so. The long-term effects of a cloud that slowly thins out could include late spring and premature fall frosts, which would impose chronic stresses on agriculture that could hamper any efforts of survivors to recover from calamity. Yet, encouragingly (sort of), the two scientists also concluded that the chances that nuclear winter could cause the extinction of humankind are vanishingly unlikely.

The NCAR study, along with others done by scientists at the Livermore and Los Alamos National laboratories, doesn't alter the possible range of climate effects of a nuclear war, from nearly none to total disaster, but it does alter the probabilities. In the TTAPS model it looked as if the climate system was vulnerable to a fairly simple on-off switch: Put just enough smoke up, and the entire system would shift into a nuclear-winter mode long enough to cause massive, possibly complete, disruption of biological functioning on earth. In the more complex models the picture becomes more diffuse. More weapons make survival more difficult—more quick freezes, more smoke, some change in the duration of acute climate change—but there does not appear to be a point of total collapse, nor does the violence of nuclear war, incredible though it is, appear to be sufficient to transform completely global climates, turning summer into winter.

This finding has lead to a debate that would be farcical if the numbers being tossed back and forth were not tallies of human deaths. Sagan has argued that extinction—or at least an extermination that would come perilously close to claiming everyone on earth—could still be the outcome of a combination of climate effects that he believes would be worse than the Schneider-Thompson projections. Schneider and Thompson respond, with their modeling results to back them up, that not everyone will die just because of climate change. The major battle, though, concerns the threshold, and that is not a scientific dispute. If Sagan is right, we have a compelling case for drastic arms reductions. What is the point, after all, of owning weapons in numbers that, if they are ever used, will only assure our own destruction?

But if Sagan is wrong, as, for the time being, it appears that he is, then the military can go back to business as usual, at least as far as nuclear winter is concerned. Here is the farce: If a global nuclear war that kills "only" two billion people or three or four is not a dreadful enough prospect to bring about some degree of prudence, then it seems highly unlikely that the threat of killing the remainder is going to bring about any startling change in policy or behavior. But even if behavior would change, nuclear winter alone no longer seems to compel such a new course. It joins the catalogue of horrors that would seem on their own reason enough not to launch a war, but the threat of climate change by itself is no magic wand. In the absence of a guaranteed threshold for disaster, nuclear winter cannot provide any overwhelming new reason that compels the nuclear powers to reduce their arsenals. Nineveh might have had an either-or choice; we do not.

No god out of the machine; nuclear winter does not deliver us from evil. We know now more than we knew five years ago. We know that a nuclear war will have long-term climate effects; we know that there are, almost certainly, long-term threats about which we remain absolutely ignorant. These may, should the worst occur, cause enormous damage in their own right. We are reminded once again that nuclear war is a terrible idea, one to be avoided at any cost—but that is all.

"I can call spirits from the vasty deep," says Glendower to Hotspur. "Why, so can I, or so can any man; but will they come when you do call for them?" responds Hotspur. Spirits still do not answer on demand. The story of nuclear winter, its rise and fall as a version of global apocalypse (and hence as global deliverance), captures the essence of the problems of scientific myths. Each increment of knowledge tells us more of our world, of the hazard in which we live. We know about the slow danger of the greenhouse effect; we know about the swift deaths that nuclear war may bring. But such knowledge only leaves us with a dampeningly mild admonition: "And now you know a little more, so act as best you can." The challenge we face is to reduce the danger of nuclear war, but this is a task for which science gives no prescription. Science can offer no certainty, no one answer, no compellingly obvious way out of a world in which nuclear war is a daily possibility. Sagan wishes it would, so does Schneider, so do I, so would anyone, but it does not. We seek from science what we cannot get—a way out of our troubles, an easy solution, a gimmick.

What we are given instead is a kind of mirror, or a telescope that twists and points inward. The nuclear-winter story provides one of the triumphs of climate science, this young science. It is a triumph to have come up with the pan-subject, the worldview, that enabled the scientists involved to pose the question that would illuminate the central issue of our day, of what we are actually capable of doing to ourselves and our world. It is equally a triumph to have begun to answer it, with the full armory of technology and models and historical analogy and planetary observations and all the details that cumulatively make up, not the science, but the grist for the scientist who can assemble the picture whole.

For nuclear winter, read acid rain, or the greenhouse effect, or global rainfall patterns, or the likelihood of excessive storminess this season or next. The success of the science has been to recognize the connections between place and place, time and time, people and all the natural world. But we do not gain along with that recognition any obvious methods of remaking connections lost or broken. Even with increased knowledge—especially with greater knowledge—the science leaves us at best with the realization that there is no simple device out there with which we can tinker to change the consequences of any particular human action. It is still up to us, not to any combination of machines and inventive systems of thought, nor to any magical Star Wars shield, to find ways to escape nuclear war, to find ways to accommodate ourselves to those changes in climate we cannot avoid. Nuclear winter is the extreme case—science has made it clear that nuclear war is even less desirable than it might have seemed a few years back—but it still establishes the paradigm: Science can alert us to an issue, but the issue itself remains for us to resolve.

So, in the end, what is the value of this change in science, this revolution in our picture of the world? Ultimately, what we get from science now (in another legacy of Galileo) is the chance to bring order—not an answer—out of the chaos of a world transformed at every turn. That order is the product of imagination, of the ability both to see into the detail (like Galileo recognizing that Venus had phases, just like the moon) and to recognize the larger whole ("it moves"). We look to science with our mixture of fear (that we will all freeze, that we are not the center of the universe) and hunger (What can keep us warm? Where are we?) because it simultaneously unsettles us and provides the tales that organize our experience, make it intelligible.

Galileo said, or we believe him to have said, "But still it moves." We cannot imagine our world fixed in place, to this day. Within the computer, we ask what will happen on the sixth day and on the seventh after a war; we ask what will happen with another fifty years' worth of carbon dioxide rising into the sky; we ask where are the ties that bind us to an entire world. We cannot now imagine ourselves unscarred by the consequences of what we do. Our telescope has turned and focuses today on a world made whole. We live now within our world, not astride it.

Further Reading

1. Boyle, R. H. and Boyle, R. A. *Acid Rain* 1983 New York, Schocken Books. A passionately argued book which makes a compelling case that (1) acid rain is a devastating problem and (2) the policymakers who have allowed the problem to reach its current proportions are culpable.

2. Ehrlich, P., Sagan, C., Kennedy, D., and Roberts, W. O. *The Cold and the Dark* 1984 New York, Norton. The first statement of the nuclear winter hypothesis. Most of the rest of the debate has taken place in journal exchanges—see especially Schneider and Thompson's argument in *Foreign Affairs* Summer 1986 and the correspondence that followed in the Fall 1986 issue.

3. Frakes, L. A. *Climates Through Geological Time* 1979 Amsterdam, Elsevier. This is an excellent introduction to the broad sweep of climate history.

4. Hansen, J. E. and Takahashi, T., eds. *Climate Processes and Climate Sensitivity* 1984 Washington, D.C., American Geophysical Union. Another of the AGU's very useful series of collections of recent research.

5. Imbrie, J. and Imbrie, K. P. *Ice Ages* 1979 Short Hills, Enslow. A good, simply written account of the development of the theory of ice ages.

6. Lamb, H. *Climate: Present, Past and Future* 1977 London, Methuen. One of the major studies of climate and human history.

7.   Langway, C. C., Oeschger, H., and Dansgaard W., eds. *Greenland Ice Core* 1985 Washington, D.C., American Geophysical Union. The sourcebook for an understanding of research on ice and the reconstruction of climate history.

8.   Lorenz, E. *The Nature and Theory of the General Circulation of the Atmosphere* 1967 Geneva, World Meteorological Organization. The essential story, told by one of the founding theorists in the field.

9.   Lovelock, J. E. *Gaia* 1979 Oxford, Oxford University Press. Lovelock's lay account of his reasoning on the Gaia hypothesis.

10.   Schneider, S. H. and Londer, R. *The Coevolution of Climate and Life* 1984 San Francisco, Sierra Club Books. A sprawling book with at least a little information about almost every topic in climate science—a book to consult with a specific question more than one to read cover to cover.

11.   Sundquist, E. T. and Broecker, W. S., eds. *The Carbon Cycle and Atmospheric CO_2* 1985 Washington, D.C., American Geophysical Union. The most comprehensive set of papers within a single binding available on carbon and its role in earth history from the beginning to the future.

12.   Wigley, T. M. L., Ingram, M. J., and Farmer, G., eds. *Climate and History* 1981 Cambridge, Cambridge University Press. A great collection of papers that examines the connections between climate and human affairs in a number of contexts.

13.   Worster, D. *The Dust Bowl* 1979 New York, Oxford University Press. The best short history of the dust bowl, illustrated with harrowing photographs.

14.   ———, *Desertification in the Sahelian and Sudanian Zones of West Africa* 1985 Washington, D.C., The World Bank. A brief introduction to the problem of desertification.

A list of technical papers on climate science published in any single year would itself fill a book. A subset of all the papers written that has played an important historical role in the development of the field is also of considerable size; a short selection of my personal favorites follows.

15. Alvarez, L. W., Alvarez W., Asaro F., and Michel H. V. "Extraterrestrial cause for the Cretaceous-Tertiary extinction" 1980 *Science* 280, published by the American Association of the Advancement of Science. The paper that began the entire debate over the idea of extraterrestrial objects having an impact on the earth's climate and evolutionary history.

16. Bjerknes, J. "Atmospheric teleconnections from the equatorial Pacific" 1969 *Monthly Weather Review* 97. The first modern statement of the argument that large-scale connections between the oceans, the atmosphere, and season-season weather patterns exist.

17. Crutzen, P. J. and Birks, J. W. "The atmosphere after a nuclear war: Twilight at noon" 1982 *Ambio* 11. The first nuclear winter paper.

18. Lorenz, E. N., "Deterministic nonperiodic flow" 1963 *Journal Atmospheric Science* 20. An abstruse and complicated paper that is the landmark in the understanding of prediction problems.

19. Neftel, A., Oeschger, H., Schwander, J., Stauffer, B., and Zumbrunn, R. "Ice core sample measurements give atmospheric content during the past 40,000 years" 1982 *Nature* 295. A report of direct measurements of the content of ancient atmospheres—a remarkable piece of work.

20. Richardson, L. F. *Weather Prediction by Numerical Process* 1922 Cambridge, Cambridge University Press. A starting point for the modern science of weather prediction.

Index